# ROBOTS DEVELOPMENT FINAL STAGE

JOHN LOK

ISBN 979-888606068-3

# Contents

# Preface

Introduction

What is future (AI) artificial intelligent products development trend and reasonable development stages? How to predict consumer behaviors to persuade who to feel (AI) products are more satisfactory to their needs? Why do consumers feel them to need to buy any (AI) products to use? Will it have other similar products to replace (AI) any products? How is the reasonable stages to achieve future (AI) development in success? If next ten year, it has not another new technological invention to impact artificial intelligence application, whether which aspects, AI is worth to invest in order to be applied to different market.

In this book, I shall give actual data to predict what the future (AI) products development trend is. Giving my opinions to predict how (AI) consumers' choices are more absolutely. In the (AI) past first stage, I shall concern travel, education, transportation, financial , hospital, administrative service etc. different job natures to indicate how to apply (AI) products to assist these industries more beneficial. In the (AI) nowadays second stage, I shall concentrate on how (AI) developing on education aspect. In the (AI) future third stage, I shall explain why (AI) will have possible to invent (AI) brain technology, even it will bring (AI) war occurence in possible.

In part one, predicting AI future 10 years three stage.In the first (AI) stage, it concerns to be given my opinions to explain how artificial intelligent technology will impact our life and will influence economic development in the future as well as how to influence human job market change. In (AI) labor market stage, I shall indicate how artificial intelligence technology influences future macro global economy change. In part two, explaining how (AI) can be applied to predict travellers' entertainment choice behaviors. In third part aims to let readers to analyse whether future (AI) artificial intelligent machine can assist humans to learn how to reduce food waste or food loss to consumers' food habits or manufacturers' manufacturing processed in order to avoid the actual food or energy shortage challenges are caused from human's food or energy consumption negligent behaviours. Consequently, I hope my readers can make judgement whether future (AI) artificial intelligence can help human to reduce or avoid environmental pollution or food/ energy waste challenges to cause social cost to be raised to influence human future living quality to be poor.

Advances in artificial intelligence (AI) technology is for the progress in critical areas, such as health, education, energy, economy inclusion, social welfare and the environment. Thus, it brings this question: Which (AI) workers be instead of traditional human workers in these different new markets? In recent years, machines had been used to be human's tasks in the performance of certain tasks related to intelligence , such as aspects of image recognition. Experts also forecast that rapid progress in the field of specialized artificial intelligence will continue. Then, it also brings this question: Does (AI) exceed that of human performance on more and more tasks? If it is truth, will some of human jobs to be disappeared? (AI) will be instead of human some simple jobs, then unemployment rate to the low skillful and low educated workers will be increased.

Whether (AI) will be raised either production or performance or unemployment to bring human job market more advantages or more disadvantages? In my this book, I shall explain whether (AI) will bring benefits or disadvantages to human job market. I shall give example to let my readers to think how to support my final view point.

What is future (AI) artificial intelligent products development trend? How to predict consumer behaviors to persuade who to feel (AI) products are more satisfactory to their needs? Why do consumers feel them to need to buy any (AI) products to use? Will it have other similar products to replace (AI) any products?

In (AI) second stage, I shall indicate how (AI) is developed on education aspect, artificial intelligent technology and online technology and online book stores are high technological intelligent product. Hence, human ourselves will have possible to cause artificial intelligent machine men to own human's mind to learn how to read books and/or write books abilities. When artificial intelligent machine men can learn how to read books and/or write books. Consequently, it means that artificial intelligent machine men can own human mind to do any jobs.

In (AI) second stage, I shall assume when artificial intelligent machine men can learn how to write books and/or read books. Then, they will have human's mind ability in possible. Can future artificial intelligent machine men be invented to learn how to write books and/or read books ability? In my this book, I shall attempt to answer this question. Finally, I hope my readers can attempt to make judgement whether artificial intelligent machine men can really learn how to write and/or books. I shall apply online technology to answer this answer. Finally, I shall give my opinions what are the influences when AI machine men had invented to achieve owning human's mind and judgement ability to our future society.

In (AI) third stage development, artificial intelligence (AI) technology is popular to be applied to different industry aspects, such as medical, construction, transportation, hospital, education etc. Although, (AI) is a human invention new development. IN fact, it seems only beneficial to human's daily life. But, it will also have threats to influence human's safety in possible , if some scientists or self-interest mind people who aim to apply (AI) to earn more profit or apply (AI) tools to be weapon to attack other countries to achieve to dominate all human's ambitious intention. Thus, (AI) will bring negative influences to our society, instead of positive influences if we can not apply this kind of new technological tools immorally.

In this stage, I shall give my opinions to indicate what reasons will cause (AI) artificial intelligent tools to be applied to social military defense weapon by human's intention. In my this books, I hope my readers can know what will cause human's immoral behaviors to bring our societies to bring more dangerous or risks or threats if human applied (AI) technology to achieve whose immoral or ambitious intention. Finally, I hope that human ought not apply (AI) technology to do any behavioral attack to satisfy ourselves interest or dominate global world ambition to avoid (AI) technological war occurrence in the future one day.

In (AI) final stage, I shall give some university lecturers' personal analytical mind to judge whether what will be occurred if artificial intelligence could be invented to match to own human brain's mind ability? What will be the advantages and/or disadvantages if (AI) robots would be invented to match to own human brain's mind ability?

I shall follow current (AI) technological development to judge whether what the potential abilities are that (AI) will achieve to satisfy human's life needs when (AI) is invented to own human brain in future one day. The main central point is discussed that whether what effects that it will bring to influence our lives when (AI) is invented to match human brain.

Advances in artificial intelligence (AI) technology is for the progress in critical areas, such as health, education, energy, economy inclusion, social welfare and the environment. Whether AI can bring positive or negative impaction to influence human job nature change.

Thus, it brings this question: Whether (AI) robotic workers can be instead of traditional human workers in these different new markets to bring positive or negative impaction to change human job nature change? In recent years, machines had been used to be human's tasks in the performance of certain tasks related to intelligence , such as aspects of image recognition. Experts also forecast that rapid progress in the field of specialized artificial intelligence will continue. Then, it also brings this question: Does (AI) exceed that of human performance on more and more tasks to replace human jobs? If it is truth, will some of human jobs to be disappeared? (AI) will be instead of human some simple jobs, then unemployment rate to the low skillful and low educated workers will be increased.

Whether (AI) will be raised either production or performance or unemployment to bring human job market more advantages or more disadvantages? In my this book, I shall explain whether (AI) will bring benefits or disadvantages to human job market. I shall give example to let my readers to think how to support my final view point.

AI future business development life cycle stage, it ought concentrate on researching how to assist office workers to work in working work environmet. Can robtics workers bring what positive or negative influences to office working environment ?Artificial intelligence had been developed to be applied to any service , working aspects, e.g. auto-driven cars. In the future, it may replace manual transport drivers in possible, e.g. non-manual driven tram, train, bus, taxi etc. public transport service. In factory warehouse environment, it can replace some workers to deliver any goods in warehouse. In shopping center service environment, it can replace security to do patrol jobs. In restaurant, it can replace waiter to deliver food to client's table. So, AI will assist or replace any service workers to do any kinds of simple jobs in any working environment in possible future.

In my this book, I shall concenrate on discussing whether artificial intelligence can bring what advantages or disadvantages to any kinds of office environment. How can they assist office workers‘ tasks ? Why employers need artificial intelligence either assist or replace any office workers' tasks? Can AI only bring advantages to office working environment ? I shall indicate different kinds of industry office environment whether AI can bring only advantages or advantages and disadvantages to any kinds of office working envirnments. Readers can have more clear understanding whether AI is real suitable technologic tool to assist any office workers' tasks.

This book divides two parts. Part one explains how (AI) brings negative impaction to influence low skillfull labor unemployment number raises , due to human job will disappear. Part one explains how (AI) brings positive impaction to influence productivity and efficiency raising , even assisting any country itself economic growth.

This book is suitable to any students who expect to make personal judgement and analysis to concern whether how (AI) technology will impact our life and future robotic business development will ought how to develop to reach mature life cycle stage.

# Prologue

Table of contents

● What is business model?
● What does business climate development strategy ?
● How to implement successful organizational downsizing strategy?
● Business growth strategy
● How whether what obstacles can affect small business growth?
● What can impact on growth strategies on business?
● Why and how can organizational life cycle models influence organizational performance?
● What is the essential elements of life cycle model to assist business development ?
● What can learn from the organizational life cycle theory?
● How to develop organizations in growth stage?
Learning organizational life cycle stage strategies
advantages
● Why do organizations need to spend time to learn how may experience different business life cycle stages?
● What advantages may bring to the organization if it can attempt to learn how to solve different challenges in different business life cycle stages ?
● What advantages to the organization, if it can know how to experience every business life cycle stage?
● The relationship between learning change management and rapid reaching mature life cycle
● How to achieve the experience of mature life cycle reaching stage rapidly for product and service ?

Chapter Nine

Robotic future final mature life cycle development stage
Must Developed And Developing Countries
Need Artificial Intelligent To Replace Human Job
● How AI help developing countries to communication and agriculture and learning and medical delivery development
p.354-370
● Emergency Response to developing countries' earthquake natural damage suddence occurrence predicting
● Smart AI Agriculture
● Medicine Delivery to developing countries' patients urgent need
● Assistance to reduce teaching work workload or psychological pressure to teachers in developing countries' schools
● Why does smart phone help developing countries communication ?

● AI can replace counter cashier service staffs

● AI can replace accountants in accountancy service industry

Why Developed And Developing Countries Need Artificial Intelligent To Assist Office Tasks

● How AI help developing countries to communication and agriculture and learning and medical delivery development

● Emergency Response to developing countries' earthquake natural damage suddence occurrence predicting

● Smart AI Agriculture

● Medicine Delivery to developing countries' patients urgent need

● Assistance to reduce teaching work workload or psychological pressure to teachers in developing countries' schools

● Why does smart phone help developing countries communication ?

● The Positive Impact of Mass Media in Developing Countries

● Why do developed countries need to develop AI

● AI may bring what benefits to developed countries

● How can AI be dangerous when developed countries continue to develop AI to become weapon to replace soldiers?

● Why the recent interest in AI safety ?

● Why do developed countries people need AI ?

Artificial Intelligence Worker Brings What Working Environment Influences

● Advantages to robotic bring to working environment

● Disadvantages to robotic bring to working environment

# (AI) Development three stages

CHAPTER ONE

# AI industry development first stage

Although, (AI) technology will be popular to applied to different jobs, but it still needs social acceptance to replace some human jobs. Today, it is increasingly common for people to use robots in various situations at home and in retail stores, hotels and hospitals. Robots are classified into several types based on their functionality ( service and utility robots or those designed to communicate with humans ) and appearance ( humanoid robots or mechanical robots). The types of robot to which every country attaches particular important in the advance of robotics, reflects the sense of values and preferences of its population . Thus, (AI) will be applied to replace human to do these above different kinds of job nature. For example, U.S. has the highest level of robot utilization at home and an retail stores with its people being the most enthusiastic about the future use of robots. Otherwise, Germany shows a strong tendency to consider robots for industrial purposes, and its people feel strong to the presence of robots in their households. Japanese accepts to apply" human aid robot" that can communicate with humans and they have a high level of familiarity with robots.

Hence, it implied those three countries have accept (AI) to replace human to do any these kinds of job duty and it will influence these three countries' workers lose their old occupations and who will unemployed absolutely, due to many (AI) robots replace them to do their job duties in the future. Also, US will have many retail service workers or retail warehouse workers are unemployed. Germany will have many manufacturing industry's workers are unemployed. Japanese will have many communication industry workers are unemployed, such as telephone service, shopping center services etc. different kind of service industry's service staffs . It will cause these kind of workers' competitive abilities are lost in themselves countries‘ jobs that require such skills include software developers, court judges, nurses, high school teachers, dentists and university lecturers, these occupations are still difficult to be replaced by (AI) robots.

Are robots taking our jobs or making them? In fact, our societies will have unemployment challenges, even (AI) technology has not created before. However, after (AI) robots invention, some of human jobs will be replaced and it can raise many low skillful and low knowledge level worker unemployment number. However, I think that high productivity driven by increasingly powerful IT -enabled machines is the causes of global labor market problems and accelerating technological change will only make those problems worse.

IT technology brings this question: Are robots killing human's jobs or benefiting human's jobs? I suppose that there is a limited amount of labor to be done. The implication is that technology can create unemployment by displacing workers, such as (AI) invention, because the more efficiently worker work ( using machines or (AI) robots), the loss work there is for workers to do. Even, any new jobs will be better done by machines or (AI) robots, and unemployment will still skyrocket. How do we know that humans will always be better at some work, or more importantly, enough work, than machines or (AI) robots, e.g. human drivers drive more safe or careful to compare (AI) robot drivers. But, the challenge is that it is not ensure that (AI) robots drivers must not drive careless to cause the chance of accident occurrences more than human drivers. However, technological change can be beneficial to innovation, automation and increasing productivity for businesses.

Consequently , it may seem machines can hurt wages and job for low skillful, less educated workers. Also, high educated workers are likely as less educated workers to find themselves displaced and devalued, and more education may create as many problems as it solves. Thus, in negative influence, automation effects on particular jobs shift workers to other jobs that are equally or more desirable. Workers may be highly compensated for possessing human capital that is specialized to a labor market. If technological advance is very rapid, such as (AI) invention, causing a large and very rapid drop in demand in a large labor market, the economy may not be able to absorb the sudden surplus of labor in a short period of timer when (AI) robots are popular to replace some workers to do some occupations in global societies.

For example, self-driving vehicles threaten to send truck drivers to the unemployment office. Computer programs can now write journalistic accounts of sporting events and stock price movement. There are even computers that can grade essay revolutionize some part of teaching jobs. Hence, (AI) robots will have possible to replace human brain to do any judgement, argument, and mind job duties. It implies some occupations which need human' mind will be threaten by (AI) robots, e.g. author, accountant, nurse, engineer. Thus, (AI) robots will have possible to replace some professional and high educated workers' jobs in the future.

But, technology can create new nature of jobs in possible. For example, a 60 minutes program indicated technology is putting new categories of jobs in the sites ( sic) of automation, the 60% of the workforce that makes its living gathering and analyzing information. Also, recession: technology kills middle -class jobs that overall technology is eliminating for more jobs than it is creating by (AI) technology. Hence, human's brain work may be assisted by 60% of (AI) gathering and analyzing information for some occupation , e.g. space scientists, ocean scientists, earth scientists etc.

However, I believe the (AI) invention and human job competition may influence global productivity change. Productivity is economic output per unit of input, the unit of on input can be labor hours( labor productivity), but if (AI) robots replace human job, then the unit of input may be (AI) machine hours ( AI) robot productivity or all production factors including labors, machines and energy ( total factor of productivity). Producing more output with less input can take several forms.

The traditional notion of productivity is a form reorganizing production and/or using better or more technology to produce more output per worker hour. But when (AI) robots invention, the form can be reorganizing production and/or using better or more (AI) robots to produce more output per (AI) robot hour. Hence, if the firm apply (AI) robots to produce its products. Then , productivity improvements in the firm may result in less workers employment, due to (AI) robots replace more worker number to achieve more productivity improvement, it has economic benefits ( less factor of production) , but more production in long term.

Thus, (AI) robots can help any firm to achieve productivity improvement in long term, for example, if unproductve farmers move to the city and start working for high-tech. manufacturers. The shift effect can be more dynamic and disruptive as low-productivity industries lose out in the marketplace to high -productivity industries and the compositional mix of the economy changes. Thus, in the long term (AI) robots can also be beneficial to high productivity industries to bring the mix of economy positive changes.

Moreover, automation will also produce some new jobs in firms that sell the new robot or other labor-saving technology. This means that, in general, there will be shift in the economy in the direction of higher-skill and higher wage jobs. Even if the (AI) robot invention country, US becomes a leader in (AI) robots producing productivity-enhancing technology, it will experience a growth in jobs serving foreign (AI) robots product buyers. Hence, (AI) robots can also create (AI) salespeople, (AI) manufacturing workers , (AI) inventors, scientists, (AI) software designer etc. occupations, when if all society does is move workers from insurance firms, restaurants and car factories to robot factories, productivity will have remained the same to create job needs for insurance, restaurant and car manufacturing worker service occupations for (AI) software designer, (AI) service robots manufacturer, (AI) service robot seller etc. related (AI) service robot product occupation created in (AI) robot technology job market. Hence, (AI) invention also create new (AI) technology job chance. (AI) impacts management job market.

In future, organization management will be changed from (AI) introduction. Division of labor will change and collaboration among humans and machines will increase. Companies will have to adapt their training, performance and talent acquisition strategies to account for a new found emphasis on work that hinges on human
judgement and skills, including experimentation and colloboration.

How (AI) impacts any organizational administrative management work? (AI) 's greatest impact will be on administrative coordination and control tasks, such as scheduling, resource allocation and reporting, (AI)-driven will place a higher premium on what we call " judgement work", the application of human experience and expertise to critical business decisions and practices when information available is insufficient to suggest a successful course of action. This kind of work will require new skills and mindsets; replacing people with machines is not goal in itself. When, artificial intelligence enables cost-cutting automation of routine work, it also empowers value -adding augmentation of human capabilities.

Thus, administrative and routine tasks, such as scheduling, allocation of resources, and reporting, will within intelligent machines, responsibilities that have long been reserved for humans. For instance, a typical store manager or a lead nurse at a nursing home must constantly juggle shift schedules, accounting for staff members' absense owing to illness, vaction, time or sudden departures. Many of these tasks will be automated by (AI). Imagine (AI) writing management monthly reports, it is not a distant dream. Leading news providers and Wall street banks are now using (AI) report generators to write news and analytical reports by drawing on quantitative data. The associated press, for example, expanded its quarterly earnings reporting from approximately 300 companies to nearly 3,000 with the help of (AI) powered software robots, freeing up journalists to conduct more investigative and interpretive reporting. For another example, Jobalime, a job-placement site, uses intelligent voile analysis algorithms to evaluate job applicants. The algorithm assesses paralinguistic elements of speech, such as tone and inflection, products which emotions a specific voice will elicit, and identifies the type of work at which an applicant will likely excel. In the future , (AI) machines can be applied to assist some kind of office administrative jobs duties. It's attractive to office managers to achieve more accurate judgment to do any administrative matters when who can be assisted from (AI) machines. Thus, managers need to spend time to learn how to apply (AI) machine to assist them to do more accurate judgement, and better informed choices. (AI) robots can be applied to improve the speed quality and cost of available products and services, instead of applying on productivity improvement and administrative improvement aspects. Thus, they may also displace large numbers of workers. This, possibility challenges the traditional benefits model of trying health care and retirement savings to jobs.

In an economy that employs dramatically fewer workers to deliver benefits to displaced workers. For example, the worldwide number of industrial robots has increased rapidly over the past few years. The fall prices of robots, which can operate all day without interruption, make them cost- competitive with human workers. In special consideration, in the service sector, computer algorithums can execute stock trades in a fraction of a second, much faster than any human. As those technologies become cheaper, more capable, and more widespread, they will find even more applicants in an economy.

Consequently, (AI) technology brings unemployed number increasing many businesses continued automating their operations rather than hiring additional workers. A trend among technology companies that receive massive valuations with relatively few workers. For example, in 2014 year Google was valued at $370 billion with only 55,000 employees, a tenth the size of AT & T's workforce in the 1960 year. Hence, if automation technologies like robots and artificial intelligence make jobs less secure in the future, there needs to be a way to deliver benefits outside of employment " flexi security" or flexible security is one idea for providing healthcare, education and housing assistance whether or not someone is formally employed.

In conclusion, (AI) and robots technology will raise unemployment to some occupations when (AI) replaces same industries' workers job duties in our societies in the future, but it also create new jobs to raise employment in any related (AI) robots and automated machine products in (AI) manufacturing. (AI) design, (AI) sale self-related industry, when (AI) replaces same industries' workers' job duties.

1.1 (AI) journalism, media publishing, digital communication technology trend

How to apply (AI) technology in digital communication journalism media, publishing industry? Some scientists indicate future (AI) and digital technology may consist such as: voice driven assistants, emerge. For example, Amazon e book publish applying digital technology and (AI) auto printing technology to sell e books to let readers to listen any e book content by (AI) voice driven speaker when they turn on computer to read e book contents; capable phones start to unlock the possibilities of 3D image of mobile story telling. New smart wearables include ear buds that handle instant translation and glasses that talk and hear. China and India will become a key focus for digital growth with innovations around payment online identity, and artificial intelligence. Thus, future (AI) technology can be applied to 3D image mobile story telling, online payment method to dealt online transaction publishing industry.

Thus, future (AI) technology can be applied to online e book publishing industry to make sound books to let readers feel more attractive . Such as Amazon publish has published sound e books to attract readers to choose to read any its books from online. Also, (AI) technology can also be applied to communication industry. For example, some online pure-play news, opinion and entertainment websites. It is a digital communication media, e.g. online journalism blog (AI) technology can be applied to visual storytellers to let online book readers to enjoy to listen to watch and send any online electronic book contents more attractive. Thus, future (AI) technology will be popular to assist any electronic book publishers to publish visual and sound talking storybook to let readers who can watch motive image and listen and read words from e books more attractive.

Thus, (AI) technology can be applied to internet ecommerce publishing or media industry to help any electronic book publishers to publish sound, image motion electronic book to attract global readers to read, even (AI) technology can be applied to digital entertainment industry, e.g. electronic 3D image virtual video games, computer games. It can be also applied to education industry, e.g. the first true digital native generation and are the native speakers of the digital language of computers to let student to learn different languages or translate words to compare to classroom learning more easily. It can be also applied to communication industry, e.g. (AI) mobile phone. Hence, it seems (AI) technology can be applied to publishing, communication, education , entertainment etc. different industries in the future. (AI) technology will be one kind of tool to satisfy human's daily life needs in the future and these industries has one characteristics is that they need to apply internet to operate to operate to do online business.

Thus, it has three trends of (AI) technology and internet technology need to be linked to achieve one kind of attractive technological business to satisfy client's needs. These three trends as below: All consumer trends involve the internet. It will be many consumer's online habits, shopping, working, socializing, watching TV, studying, travelling, listening. Thus, (AI) music, eating and exercising are just a few examples. This is happening because human usually use mobile broadband or Wi-Fi, rather than cables. Thus, (AI) technology will be applied to mobile to satisfy client's need absolutely.

The mobile phone can be more popular to be used more than computer or laptop tools. The reasons are because women dive the smartphone market by defining mass-market use. But as the speed of technology adoption increases mass market use becomes much quicker then before. Successful new technological products and services , such a (AI) mobile phone products now reach the mass market in popular use. It means that the time period when early adopters influence others is shorter than before. Also, since new products and services increasingly use the internet mass markets are not only faster , but are also more important than ever to consumer themselves. Most internet services become more valuable to individuals when many use them. Thus, it causes why (AI) mobile phone will be popular to be used.

Since, new products and services increasingly use the internet, in the future several trends focus on (AI) smart phone users. Consumers' familiarity with using smartphone apps. Essentially, the technologies will bring other related (AI) and internet service needs, e.g. sound and image emotion e book needs, (AI) mobile communication needs, e-virtual games or e-3D image virtual games etc. entertainment activities needs with such a large part of the world's population now online, it is clear that there is strength in numbers.

Thus, (AI) imagines , if future any (AI) and internet related services or products new technology is easy to use and inexpensive, when the latest products reach the mass market almost as quickly as they reach the early adopters and industry experts. I believe that any (AI) and internet related products or services must be popular to accept to consume for entertainment or useful aim. For example, with major players including Apply, Facebook and Google had invested (AI) technology to develop their businesses. (AI) technology has the potential to disrupt everything in the coming years, from the lives of connected consumers to every industry (AI) will be an alternative route for brands to reach consumers with convincing and relevant messages. Digital technology will assist of the future, then it can improve technology to bring this effect, such as sophisticated software machine learning and speech recognition effective. Hence, Google, Facebook , Yahoo web site service companies can apply (AI) technology to help other companies to advertise their businesses, such as travel, retail, and education etc. industries more attractive. (AI) technology can be applied to internet company to be aware and familiar enough to drive among mainstream consumers, it can create online experience to travel, retail , education and other entertainment needs to online consumers to seek their entertainment needs more easily. Hence, in the future (AI) technology and internet related entertainment service needs will be raised in this (AI) and online consumption market.

1.2 (AI) healthcare service industry development

In the future, (AI) medical internet technology tool can be applied to assist individual's health at the center of their focus, e.g. smartwatch compatible mobile app. patients can let personalized reminders for taking their medication snap pictures of their prescriptions to expedite refills, and scan their insurance card. So that, store clerks are prepared with up-to-date patients' information . (AI) owned health operated technological clinics can help patients to receive treatment for minor illnesses, flu shots, cholesterol screenings and more than a dozen other medical services, all of which can be patients who can't make it to a physical location. (AI) healthcare services organizations can provide various telemedicine services. So, patients can receive care via phone or video chat.

For example, one London-based intelligent Brewing company has developed an (AI) system to continuously collect and incorporate customer feedback, which the system itself uses to brew ne various of the company's beers. Thus, the beer clients can give feedback to talk to the algorithm (AI) machine, whenever or anywhere who're drinking the beer. It is such any healthcare services organizations can apply (AI) machine to collect patient's feedback to talk to the algorithum (AI) machine whenever or anywhere who're eating any medicines. So doctors can know every patient's health conditions any time. If the patients feel uncomfortable, the doctor can know from (AI) machine notification to decide whether the patient needs to eat another new medicine or keep to eat same medicine is better. Hence, (AI) medial internet technological body check report machine will be proper to be needed to serve any hospitals' patients in the future.

However , it brings this question. How can (AI) medical internet technological body check report machine apply to hospital more efficient? The essential new medicine co-workers for the health service digital age health service leaders need apply (AI) medical report machines and artificial intelligence to the newest recruits to the workforce bringing new skills to help health service staffs do new jobs and reinventing what's possible, building the health service workforce for today's digital health service demands for patients. Thus, technology-driven health service model innovation from the health service organization outside in and providing digital health service ecosystems for patients to use the (AI) health service equipment will be popular to be accepted to be used.

1.3 Robot and internet things future machine men invention

Nowadays, there are some company, which apply internet and (AI) I Robot technology to do any similar human job nature. For fishing industry example, one company, known for creating the Roomba, I Robot is now working with marine conservationists to launch an ocean-patrolling intelligent robot to hunt and manage invasive species, protecting native populations. And evolved industries like precision agriculture are ramping of our increasing population. Area of practice that once seemed impossible to digitize are fundamentally changing because of the impacts of (AI), internet of things capabilities and big data analytics, which have many potentially positive impactions for society.

For textile industry example, automation is nothing new, it has shaped the workplace to replace human jobs to boost productivity in the textile industry. Textile machines have had a generally positive impact over gears, creating value and allowing textile workers to take up more rewarding age will likely continue to create opportunities and lead to new textile industries, companies and textile occupations. It may also compensate for a demographically driven slowdown in the growth of the textile workforce. The future impact of textile (AI) and automatic and internet link is somewhat uncertain. It seems textile industry will be trend to accept (AI) textile workers and internet of thing to replace traditional manual textile workers to produce any shirts, cloths etc. wearing products in factories popularly, during the (AI) textile machine and internet thing technology can be invented to reach the mature stage in the future.

For factory worker transportation job example, they have also expanded their influence, migrating from the factory floor to the service sector and taking the place of humans in a range of activities from financial transactions to transport route optimization. Further (AI) machines and robots are increasingly programmed to learn, meaning they improve with time and undertake cognitive activities. Hence, (AI) machines and internet technology enable automation of work activities to raise factory workers' efficient and performances, also factories can reduce manual worker numbers, due to (AI) machine workers' assistance.

Future, (AI) robotics technologies and internet technique have these different kinds of characteristics: For soft robotics example, it is non-rigid robots construct with soft and deformable materials that can manipulate items of varying size, shape and weight with a single device. For swarm robotics, it coordinated multi-robot systems often involving large numbers of mostly physical robots. For touch/factile robotic example, it robotic body pails (often biologically inspired hands) with capability to sense, touch , dexterity robots example, serpentine robots with many internal degrees of freedom to threat through tightly packed spaces for humanoid robots example, robots physical is similar to human being often bi-pedal that investigate variety capable of performing human tasks , including movement across terrains, object recognition, speech sensing etc. For autonomous cars and trucks example, it is capable of operating with a human pilot, e.g. the unarmed general atomics Predator XPUAV with roughly half the wingspan of a Boeing 737 can fly autonomously for up to 35 hours from take-off to landing, for unmanned aerial vehicles example, flying vehicles capable of operating without a human pilot, the unarmed general atomics predator -XPUAV , with roughly half the wingspan of a Boeing 737, and fly autonomously for up to 35 hours from take off to landing, for (AI) chat bots example, (AI) systems designed to simulate conversation with human users, particularly those integrated into massaging apps.

In Dec. 2015. the general service administration of the US Govt. described how it used a chat bot named Mrs. Landingham (a character from the television show the west wing) to help onboard new employees. Finally, for robotic process automation example, class of software robots that replicates the actions of a human being interacting with the user interfaces of both software systems. Enables the automation of many back-office work flows without requiring expensive IT integration . Hence, future (AI) robot machine men will have different functions to be applied to different industries to use in possible.

Statistics Denmark shows that (AI) automation potential robots will influence few jobs are completely automatable , but close to half consists of 40% automatable tasks: It showed example occupations include share of automated, such as brewing machine operators are more than 80%, logging equipment operators are more than 50%, roofers , stock tasks clerks, travel agent are more than 50%, farmers , nursing assistants are more than 30%, physicians, teachers , managers are more than 10%.

For example, humans perform a wide variety of tasks from planting corn to examine spreadsheets, meeting clients and lifting crates in a store. Each of these actions requires a combination of innate or acquired capabilities, internet technique assistance, ranging from social perceptiveness to fine motor skills and natural language understanding. To understand and map automation feasibility by existing technology. Mc Kinsey has developed a framework of 18 technical capabilities that can substitute tasks performed by humans. The capabilities are grouped in five categories: sensory, cognitive, language, social and emotional and physical. So, it seems (AI) robot machine men and internet technique will have possible combination to invent to own human's emotion , language, learning, task skill abilities.

Mckinsey global institute analysis also showed current technologies have achieved different levels of human performance across 18 capabilities include: sensory perception, autonomously infer and integrate complex input using sensors, cognitive capabilities reorganizing known patterns/ categories supervised learnings, generating novel, logical reasoning/ problem solving, optimization and planning, creative, information retrieval, coordination with multiple agents, output articulation/presentation, national language processing, social and emotional capabilities-natural language understanding, social and emotion sense, reasoning output, physical capabilities-fine motor skills, navigation mobility. Hence, it seems (AI) robots and internet technological will combine to invent to own human‘ some skills to replace human to do some kind of tasks in possible.

In conclusion, future (AI) robot and internet will be needed to link to cooperate together to raise human's work efficiency in popular.

1.4 How artificial intelligence replaces human job possibility

What is the risk of automation for jobs to replace human job? In recent years, there has been a revival of concerns that automation and digitalization night after all result in jobless future. As I argue, this might lead to an overestimation of job (AI) automate , as occupations labelled as high-risk occupations often still contain a substantial share of tasks that are hard to automate.

For example, when the share of (AI) automatable jobs is 6% in Korea, the corresponding share is 12% in Australia. Differences between countries may reflect general differences in workplace organization, differences in previous investments into (AI) automation technologies as well as differences in the education of workers across countries. I also discover that (AI) automation and digitalization are unlikely to destroy large numbers of jobs. But, however, low qualified labors are likely to raise costs as the (AI) automate of their jobs is higher compared to highly qualified workers.

In fact, (AI) technology will influence some new technology to replace some human's job, such as driverless car, the largely autonomous smart factory , service robots or 3D printing. These technologies are driven by advances in computing power, robotics and artificial intelligence and ultimately redefine what type of human capabilities machines are able to do.

Hence, question brings whether (AI) invention will influence general human jobs to be replaced by (AI) autonomous jobs? Whether will the potential foe automation with actual employment loss? In particular, the technical possibility to use (AI) machines rather tasks need not mean that the substitution of humans by machines actually takes place.

Whether (AI) technology replaces human's some job, it is beneficial to our society or not. Instead, machines are increasingly capable of performing non-routine cognitive tasks, such as driving or legal writing . In particular, advances in the field of machine learning (ML), e.g. computational statistics and visions, data mining, artificial intelligences allow for automating cognitive task, when the use of (ML) in mobile robotics (MR) also allows for automating certain manual tasks. So, it seems, (AI) technology can replace some labor job, e.g. warehouse transportation, even mind's job, e.g. legal writing, driving in possible.

For example, if (AI) automatic non-manual driving can reduce hurt or death risk, it is beneficial to our society, or (AI) automatic robots can more any heavy things ( products) in warehouse safely. Then, it can reduce the warehouse labor's bodies hour risk, it is beneficial to the workers. Even, if (AI) robot can write any legal documents, no any word errors in short time. It is beneficial to the law companies , but it also bring unemployment chance, due to these jobs can be replaced by (AI) robots to do in the future. Hence, it will cause some occupation to be disappeared, due to (AI) robots can do our these kinds of jobs in the future.

Frey & Osborne (2013) reported these kinds of occupations will be replaced by (AI) robots in possible. They include computer, engineering, financial, management, legal , art and medium, community service, education, healthcare practitioners and technical service, sales and related, office and administrative support, farming, fishing and forestry, construction and extraction, installation, maintenance, and repair , production, transportation and material moving. It seems our future some professional occupations will have possible to the replaced by (AI) robots to replace, instead of labor jobs. Hence, (AI) robots technology will have much trend to replace high knowledge or low knowledge skillful labors in the future.

In conclusion, it implies that only using information on task-usage at the individual level leads to significantly lower estimates of jobs " at risk", some workers in occupations with according to high automate nevertheless often perform tasks with are hard to automate. Why can (AI) replace human to do some kinds of jobs? (AI) artificial intelligence refers to the ability of a computer or a computer enable robotic system to process information and produce outcomes in a manner similar to the thought process of humans in learning, decision making and solving problem. By extension, the goal of (AI) systems is to develop systems to capable of tasking complex problems in ways similar to human's logic and reasons who feel in our future. Hence, it means future (AI) robots has effort to replace human to do any jobs in possible.

1.5 (AI) directions for future non-manual control road vehicles market

Future road vehicle products and technologies must meet social, economic and environmental protection and driving safety goals , and satisfying market requirements for mobility, accident reducing, performance, cost desirability. Thus, (AI) auto-non manual control vehicles need to be followed this direction to invent. To satisfy future driver's safety of needs, enhanced vehicle speed desired functional performance of road transportation system, required and desired technological response, including research needs. It is long term up to 20 years vision, for (AI) auto non-manual research. Thus, (AI) auto non manual vehicle manufacturers need often to revise their (AI) vehicles functions to raise to improve their system performance and driving industry driver's needs, e.g. private drivers need or public transportation driver's need or business client's need. Hence, future (AI) transportation will need have individual driving consumer and business driving consumer both targets.

Thus, future (AI) automation manual control vehicles need to deliver high impact technology solutions to meet social , economic and environmental and safe goals. Engine needs to be improved efficiency , performance, drivability, reliability, durability and speed-to-market together with reduced emissions and cost; hybrid, electric and alternatively fuel (AI) non manual control vehicle technology development, leading to new fuel and power systems, such as hydrogen, fuel cells and batteries, which satisfy future social, economic and environmental and safe goals. Software, sensors, electronics and telematics technology development are needed to be lead to improve vehicle performance, control and adaptability, intelligent , mobility and security, structure and materials technology development, leading to improved safety, performance and leading to flexibility with reduced cost and environmental pollution to achieve the (AI) non manual drivers to feel (AI) vehicle performance, auto control and adaptability is better to compare traditional manual driving vehicles.

In fact, in traditional manual driving market, Japan and USA had had over 80% of world car production by six major global groups. In the future, it is possible only USA can dominate (AI) non manual auto driving vehicle manufacturing market if Japan had no effort to manufacture any (AI) auto non manual control vehicles. So, it means that it is only Japan is USA potential (AI) auto non manual control vehicle manufacturing competitors. Also, it means that it is only USA has effort to export (AI) auto non manual control vehicles to global (AI) auto non manual control vehicle market.

Thus, in long term, (AI) non manual control vehicle product market development, USA (AI) vehicle manufacturers will have these requirement to win new technological competition to traditional manual control vehicle. The requirements include: low cost fuel, low carbon, fuel cell and telematics technologies, the technological roadmap function, such as detailed consideration clear provision of other important areas to the drivers. When the (AI) non manual auto drivers are sitting in the non manual control auto vehicle. Although, who does not need to drive, but who need to know how to go to anywhere by electric road map show clearly. So, the driver won't lose direction and he/she can know the (AI) non manual control vehicles is driving to anywhere in any time, even when who is sleeping.

In the future, the (AI) non manual control vehicles need to be invented to satisfy any business , transportation clients' needs, instead of individual clients needs, e.g. cans, trucks, buses, emergency and utility vehicles, trains, trams etc. Hence, technological road mapping is one important tool to help any business, transportation ( AI) non manual control vehicle clients. Technological roadmap is a technique that is used in industry to support strategic planning for (AI) non manual driving vehicles in the future. Electronic road maps generally take the form of multi-layered time based charts, linking technology developments to future (AI) non manual control vehicle market requirements.

Technology road mapping is a flexible technique and the roadmap architecture and process for developing the roadmap most generally be customized to meeting the particular aims. Why technology roadmap will be popular to (AI_ non manual control vehicles. It's advantages include: It is a technology solutions and options that can enable the performance targets to be achieved engine hybrid, electric and alternatively fueled vehicles, software, sensors electronics and telematics, structures and materials design and manufacturing process. It is road transport system performance measures and targets tool, in response to the trends and get (AI) non manual control vehicle drivers to get society, economy, environment protection, low cost driving benefits, also it can help any transportation clients to know how to go to anywhere clearly. Hence, technological road map will be one good tool to assist (AI) non manual control auto vehicle to develop future road driving market.

Reference(source)

Frey & Osborne (2013), The future of employment: How susceptible are jobs to computerization? University of Oxford.

Mckinsey Global Institute Analysis

Statistics Denmark, Global automation impact model, Makinsey analysis

1.6 (AI) -driven automation industry development

(AI) -driven automation industry will create wealth and expand economy growth to any countries, but it will be accompanied by changed in the skills that workers need to learn. One of main ways that technology increases productivity is by decreasing the number of labor hours needed to create a unit of output. It implies (AI) technology will influence low educated and low skillful labor number to be decreased ( reduction employment number).

In contrast, technological change tended to work in a different direction throughout the nowadays. The advance of computer and the internet raised the relative productivity of higher skilled workers. So, routine-intensive occupations that focused on predictable tasks disappearance, such as switch board, operators, filming checkers, travel agents and assembling line workers etc. were particularly replaced by new technologies.

However, today, it may be challenging to predict exactly which jobs will be most immediately affected by (AI) driven-automation. The reason is because (AI) is not a single technology, but rather a collection of technologies that are felt unevenly through the economy to influence job changing both negatively and positively. In positively view point, (AI) driven-automation will make many workers more productive and increase demand for certain skills. Consequently, new jobs are likely to be directly create in areas , such as the development and supervision of (AI) as well as indirectly created in a range of areas throughout the economy as higher incomes lead to expanded demand. Otherwise, in negatively view point, many traditional human needed ( demand) skillful jobs will be threatened by automation are highly concentrated among lower-paid, lower-skilled and less -educated workers. It means automation will cause pressure on demand for this group, pressure and employment, if (AI) can replace the low skilled and less educated workers' jobs. Thus, (AI) will have negative influence to impact on the labor market.

(AI) capabilities will enable automation of some tasks that have long required human labor. Why can (AI) replace some simple human jobs? For example, advances in robotics are expanding machines' abilities to interact with and sharp the physical world. Combined , (AI) and robotics will give rise to smarter machines that can perform more sophisticated functions than ever before and brings more advantages that humans have exercised. This will permit automation of many tasks now performed by human workers and could change the shape of the labor market and human activity.

1.7 How (AI) influences labor market

Today, it may be challenging to predict exactly which jobs will be most immediately affected by (AI)-driven automation. Because (AI) is not a single technology, but rather a collection of technologies that are applied to specific tasks. Some specific predictions are possible based on the current (AI) technology. For example, driving jobs and house cleaning jobs, bank counter service jobs, telephone enquiry service operators. Restaurant cooking jobs, simple accounting record service jobs etc. that require relatively less education to perform. Advancements in computer vision and related technologies have made the feasibility of fully appear more likely, potentially displacing some workers in driving-dominant professions. Seemingly similar robot, for which the operational tasks is less specific of navigating to a specific destination when following a set of given rules and preserving safety.

In the future, the effects of (AI) on the labor market in the decade ahead will continue the trend toward skill-biased change that computerization and communication innovations have driven in recent decades. Thus, some human driving occupation will be disappeared or replaced by (AI) automation driven. For example, bus drivers, light truck or delivery services drivers, heavy and tractor-trailer truck drivers, school drivers, tax drivers, travel bus drivers.

However, (AI) technology could enable some workers to focus time on other job responsibilities, boosting their productivity, and actually raised wage growth among those still holding the reshaped jobs. For example, salespeople, who currently spend a considerable amount of time driving could find themselves able to do other work when a car drives them from place to place, or inspectors and appraisers could fill out paperwork, when their car drives itself. This (AI) -driven technology should make these workers more productive, with (AI) -driven technology serving as a complement, not a substitute. New jobs will also likely be created, both in existing occupations cheaper transportation costs with lower prices and

increase demand for products and all the related occupations, such as service and fulfillment, and in new occupations not currently foreseeable.
What kind of jobs will be created by (AI) technology? Predicting future job growth is extremely difficult, due to it depends on technologies or substitute for existing today as well as they may complement or substitute for existing human skills and jobs. However, (AI) will also lead to substantial indirect job creation to the degree it raises productivity and wages, it may also lead to higher consumption that would support additional jobs from high-end draft production to restaurant and retail. The future(AI) " augmented intelligence", the technology's role is as assisting and expanding the productivity of individuals rather than replacing human work. Thus, based on the biased-technical change framework, demand for labor will likely increase the most in the areas where humans complement (AI) automation technologies. For example, (AI) technology , such as IBM's Watson may improve early detection of some cancers or other illnesses, but a human healthcare professional is needed to work with patients to understand and translate patients' symptoms, inform patients of treatment options, and guide patients through treatment plans. Shipping companies may also partner workers who pick up and deliver products over the last feet with (AI) enabled autonomous vehicles that move workers efficiently from site to site. In such cases, (AI) augments what a human is able to do and allows individuals to either be move effective in their specially task or to operate on a larger scale. Thus, it seems (AI) technology will also create new jobs, raise productivities and workers' efficiencies.

1.8 Redefining management in the workforce of artificial intelligence

In the future, due to artificial intelligence influences to some kind of human jobs nature. So, the kind of human jobs of management methods will also need to change to adapt the artificial intelligence technology input to their organizations. It will cause challenges for every executive and manager if who won't have effort to manage their teams how to apply artificial intelligence technology to work efficiently and easily. For example, division of labor will change among humans and machines will increase. Thus, companies will have to adapt their training performance and talent strategies how to emphasize on work that how to make human judgment and skills and experimentation. Thus, (IA)'s greatest impact will be on administrative coordination and control tasks, such as scheduling , resource allocation.
In fact, mangers will encounter this challenges: How to apply human

experience and expertise to judge critical business decisions and practices when the information available is insufficient to suggest a successful course of action? Due to this kind of work will require new skills and mindsets. I shall indicate these change management methods to adapt (AI) technology. Such as: administration and routine tasks, scheduling , allocation of resources and reporting will fall within the intelligence machines, responsibilities that have long been reserved for humans. For example, a typical store manager or a lead nurse at a nursing home most constantly arrange shift schedules, accounting for staff members' absences owing to illness, vacation time or sudden departures.

Thus, the managers need to learn how to arrange new division of labor within the organizations after (AI) technology had been implemented to the organization. Artificial intelligence is currently influencing into once considered exclusive to humans: assessing and acting on human emotions and personality traits. The influences to managers need to change their strategies to adapt (AI) technology implements include such as below:

Firstly, managers need to spend the bulk of their time on coordination and control tasks from intelligent system implements. Their time spending on these major three aspects from impact of intelligent system: coordinate and control, solve problems and collaborate and people and community , strategy and innovation three aspects. Thus (AI) will influence managers need to change their judgment method to teach whose teams how to adapt the (AI) system operations in any organizations.

Secondly, (AI) will influence top, middle and low level management needs to change to adapt the (AI) technology operations to any owned (AI) technology organizations in the future. Intelligent machines must be trained in context. Just like humans , on-the-job training is a requirement for such machines because they typically arrive with only very general capabilities. To get the most from (AI), managers at all levels must participate in the instructional experience and in the learning process and provides managers' familiarity with such systems on these aspects, e.g. How the system works and generate advice, how the system has a proven track record , how the system provides convincing explanations , how the system can make simple rule- based decisions.

Thirdly, managers need to learn how to make judgment more accurate (AI) systems assistance. Although (AI) will invariably take on more routine work and even augment human decision-making, it won't judgment work, the application of human experience and expertise to critical business decisions when the information available is insufficient to suggest a successful course of action or reliable enough to suggest an obvious course of action. For a sense of the nature of judgment work, consider big data marketing and sales analytics. Such analytics often provide insights that can inform promotional campaigns, including predicting which promotions will generate desired sales brand further into the future, marketing executives need use judgment, combining analytics with their own and others' insight and experience.

The application of experience and expertise to critical business decisions and practice represents the real value of human judgment. But, when artificial intelligent machines are implemented to any organizations to assist the low, middle and top level management to make any business judgment. These forms of judgment work that managers can gather data interpretation, idea development more absolute from (AI) machine assistance. Thus, why these level management executives need to learn how to apply (AI) machines to help them to make any business judgment more accurate.

1.9 How (AI) influences organizational change

Consequently creative and social intelligence will be in even greater demand as (AI) makes in management and the workforce. This development will represent a long term trend in labor markets , one characterized by intensifying demand and reward for social skills with a growing desire for creative capabilities, managers will seek to fashion of ideas and hypotheses from inside and outside of the enterprise to shape solutions to their most pressing business problems. Thus, (AI) will influence overall organizational team members who have chance to participate any decision to make more accurate business judgment.

Many managers mistakenly view judgment work as only an individual discipline, failing to appreciate that it can also involve decide interpersonal and organizational practices. In more complex settings, judgment is typically a collective outcome of individuals' and teams' diverse perspectives, insights and experiences. And often , the resulting choices are better informed than decisions that an individual would have arrived at on his or her own.

Thus, when any organizations apply (AI) technology to assist managers to gather data and ideas to make any judgment. In these cases, organizations can create the conditions for effective collective judgment by establishing structures , such as " shadow advisory boards" that prompt managers and employees to source and synthesize multiple perspectives. Thus, a traditional organization (firm) might freshen its thinking is t put together a shadow advisory board, comprised of young, digital people who can apply (AI) machine assistance to make judgment work more accurate whether related to people development, problem-solving or strategizing and innovating for considerable degrees of creative and social intelligence.

Thus, on the one hand, (AI) technology machine augmentation and automation can give these advantages to human (organization managers) , e.g. developing people and community, solving problems and collaborating, coordinating and controlling work, shaping strategy and leading innovation. Besides, on the other hand, the next generation managers need have these individual attitude to treat intelligent machines to be as colleagues.

When, judgment is a human skill, intelligent machines can accelerate human learning that supports it, assisting in data -driven simulations, scenarios and search and discovery activities. Focuses on judgment work, some decisions require insight beyond what data can tell them. This is the sweet sport for human judgment, the application of experience and expertise to critical business decisions and practices. Thus, managers will also need to find ways to learn how to use digital (AI) technologies to tap into the knowledge and judgment of partners, customer external stakeholders and role models in other industries after the (AI) machine had been implemented to the organization.

1.10 Future works change:
Automation, employment
and productivity

Human future " micro to macro" industry trends will be affected business strategy and public policy by (AI) technology. In the future (AI) technology will influence those six themes: productivity and growth, natural resources, labor markets, the evolution of global financial markets, the economic impact of technology and innovation and urbanization. However, (AI) technology will bring economic benefits of tackling gender inequality, a new global competition, Chinese innovation and digital globalization.

Nowadays, advances in robotics artificial intelligence, and machine learning are in a new age of automation, as machines match or outperform human performance in a development to any countries. For example, automation of activities can enable businesses to improve performance by reducing errors and improving quality and speed, and in some cases achieving outcomes that go beyond human capabilities. For example, some research indicated automation could raise productivity growth globally by 0.8 to 1.4 % annually; more than 2,000 work activities across 800 occupations. When less than 5% of all occupations can be automated using demonstrated technologies about 60% of all occupations have at least 30% of constituent activities that could be automated. Many occupations will change that will be automated away: Activities most susceptible to automation involve physical activities, in highly structured and predictable environments, as well as the collection and processing of data. They are most prevalent in manufacturing , accommodation and food service and retail trade and include some middle-skill jobs. For example, such as natural language processing is a key factor. Beyond technical feasibility, the cost of technology competition with labor including skills and supply and demand dynamics, performance benefits including and beyond labor cost savings, and social and regulatory acceptance will be affected by (AI) automation technology. Thus, (AI) automation will impact to influence global employment in those aspects as below:

Firstly, assuming that people are displaced by automation will find other employment. The anticipated shift in the activities in the labor force is of a similar order as the long-term shift away from agriculture and decreases in manufacturing share of employment. Both of manufacturing and agriculture industries which would be accompanied by the creation of new types of work not foreseen at the time.

Secondly, for business, the performance benefits of automation are relatively clear. Thus, the businessmen have opportunities for their micro economies to benefits from the productivity growth potential and macro economies to benefit to encourage continued progress and innovation , investment and market incentives. At the same time, employers must innovate policies to help workers and institutions adapt to the impact on employment.

This will likely include rethinking education and training, income support and safety nets , as well as support for those dislocated, when employees need to leave themselves homes to move to other cities to learn new (AI) automation works. Thus, individuals in the workplace will need to engage move comprehensively with machines as part of their everyday activities, and acquire new skills that will be in demand in the new automation age. Consequently , the scale of shifts in the labor force over many decades that automation technologies can be a similar order to the long -term technology

-enables shifts in the developed countries' workforces away from agriculture in the 21 th century. Those shifts did not result in long-term mass unemployment because they were accompanied by the creation of new types of work not foreseen at the time. However, human will still be needed in the workforce when the total productivity gains are caused by (AI) technology.

1.11 What occupations will be influenced by (AI) technology.

In the future, scientists predict that these occupations will be influenced by (AI) technology mostly. They include : retail salespeople, food and beverage service workers, language or translation teachers, health practitioners. Since these work activities have a more relevant occupations are made up of a range of activities with different potential for (AI) automation . For example, a retail salesperson will spend more time interacting with customers, stocking shelves , or ringing up sales. Each of these activities is distinct and requires different capabilities to perform successfully.

Thus, these job activities have similar simple control characteristics. Simple activities include greet customers, answer questions about products and services, clean and maintain work areas, demonstrate product feature process sales and transactions. All these activities can have similar simple activities in order to (AI) machines can be learn how to do these activities from (AI) technology . For example, the capability perception includes sensory perception, cognitive capabilities, such as retrieving automation, recognizing known patterns( supervised learning), logical reasoning problem solving.

Thus, (AI) machine is such human, which has feeling and emotion, such as social and emotional sensing, judgement reasoning methods, natural language understanding and physical capabilities, such as mobility , navigation, gross motor skill, fine motor skills. It seems that the future, (AI) human invents machines which will have these human characteristics to do human similar behavioral job duties more easily and efficiently. It implies these above human occupations will be replaced by (AI) human invention machines in the future. Due to (AI) creation, it is possible to cause unemployment number of these above workers will increase because (AI) machines can do their similar job behavioral activities.

Consequently, employers won't need to employ many of these skillful labor. Otherwise, they can buy less number (AI) machines to attempt to do whose job activities more easily and efficiently. So, it seems (AI) machines will have more high work performance to replace these occupation workers' work performance. Finally, these occupation worker unemployment number will only increase when the (AI) machines had been invented to achieve to do their work behavioral activities absolutely success in the future.

## 1.12Whether (A) technology machine labor will replace human worker more or assist human worker more

There is no single agreed definition of a robot how outcome of a task that is completed without human intervention. When some definitions require the task to be completed by a physical machine moves and respond to its environment, other definitions use the term robot in connection with tasks completed by software , without physical embodiment.

However, to answer the question : Whether (AI) technology machine labor will replace human worker more or assist human worker more. I shall indicate some examples to let readers to judge whether (AI) technology can create new jobs or reduce old jobs.

Firstly, I shall explain what (AI) function is. (AI) is a service robot that performs useful tasks for humans or equipment excluding industrial automation application . Thus, the classification of a robot into industrial robot or service robot is done according to its intended application. It is also a personal service robot or a service robot for personal used for a non commercial task, usually by lay persons . Examples are domestic servant robot, and pet exercising robot. It is also a professional service robot or a service robot for professional used for a commercial task, usually operated by a properly trained operator. Examples, are cleaning robot for public places, delivery robot in offices or hospitals, fire-fighting robot, rehabilitation robot and surgery robot in hospitals. Thus, these functions will be future (AI) application to our daily life necessaries or business necessaries.

However, some authors agree (AI) will bring negative outcomes of automation, due to raise competiveness, reduce human job nature. Otherwise, other authors argue (AI) will bring positive outcomes of automation, due to raise productivities, job creation, assist humans work.

On the positive outcome hand, robots can increase productivity . This is particularly important for small-to medium sized businesses both are in developed and developing countries economies. It also enables large companies to increase their competitiveness through faster product development and delivery. Increased use of robot is also enabling companies in high cost countries to re shore, or bring back to their domestic base parts of the supply chain that will have previously outsourced to sources of cheaper labor. Currently , the greater threat to employment is not a automation, but an inability to remain competitive. Automation has led overall to an increase in labor demand and positive impact on wages. The reason is that the middle-income/middle-skilled jobs have reduced as a proportion of overall contribution to employment and earnings leading to fears of increasing income inequality, the skills range within the middle income bracket is large. Thus, robots are driving an increase in demand for workers at the higher -skilled and with a positive impact on wages. This issue is how to enable middle-income earners in the lower-income range to unskilled or retain. Finally, the (AI) positive impact supporter who argue the future will be robots and humans can work together.

However, on the negative outcome hand, robots can substitute labor activities, but don't replace jobs. They believe that less than 10% of jobs are fully automatable. Increasingly , robots are used to complement and augment labor activities, the net impact on jobs and the quality of work is positive. Automation can provide the opportunity for humans to focus on higher-skilled, higher-quality and higher-paid tasks. Robots can improve productivity when they are applied to tasks that which perform more efficiently and to a higher and more consistent level of quality than humans. For example, increased productivity is enabling some firms, such as Whirlpool, Caterpillar and Ford Motors company in the US restructure their supply chains, bringing back parts of the manufacturing process to the country of origin. Thus, productivity gains due to robotics and automation are important not just at the company level, but also for build industry and nation competitiveness.

I suppose that productivity can be raised. What are the impacts of robots on employment? Firstly, the main focus of development has been on personal entertainment, which does not drive worker productivity ( manufacturing production). When the internet ( information and communication technology (ICT)) innovation. This is borne and by findings that manufacturing productivity, which has been driven by innovations in automation rather than consumer technologies, has government strongly than productivity in the services sectors of the economy in most nature economies. It seems (AI) automation will create many jobs in internet communication entertainment game industry. For example, many young people like to use internet to play any electronic games from computer or mobile at home or outside home conveniently. Thus, (AI) automation will increase demand to be invented to any new entertainment game from internet channel. It will need to employ many (AI) entertainment game inventors to create many automation entertainment games. Thus, (AI) automation in internet entertainment game industry will need human (AI) entertainment game inventors to invent the knowledge-based capital of (AI) automation entertainment games. The (AI) entertainment game inventors will need own research and development skills, form specific skills, organizational know-how skills, databased knowledge, design and various forms of intellectual property to do these (AI) automation entertainment game invention occupations in the future.

International Federation Of Robotics(2016) indicated that China will be as a major robotics manufacturer and user of robots, benefiting from jobs created by robot manufacturing and productivity gains from robot use. Chins had sold of robots to any one single market every year since 2017 year. The Chinese government has included a focus on robotics in its 10 year strategy. In order to achieve its target of a robot density of 150 units per 10, 000 workers by 2020 year. Thus, Chinese companies will have to install around 650,000 new industrial robots between 2016 to 2020 year, 2.5 times more than installed globally in 2015 year.

Hence, China (AI) manufacturing industry will need to employ many workers . It implies (AI) manufacturing industry will create many new occupations in China. Also, ministry of economy, trade and industry (2015) also showed that Japan currently has the largest stock of industrial robots in operations, primarily in the automation industry. Driven by a rapidly aging population and low productivity rates, the Japanese government has sights on a 20-fold increase in the use of robots in the non-manufacturing sector and a three-fold growth rate of labor productivity in the service sector both by 2020 year. Thus, it also implies Japan will need many robots to be provide to service industry. Due to robots will provide to serve any businessmen's clients. Thus, it is possible that the service workers won't be dismissed as well as it is depended on the serving job nature to decide whether Japan's service workers can still serve to their employer when the service (AI) robots are applied to whose employers.

Consequently, it seems that (AI) can create employment, Ministry of economy, trade and industry (2015) showed that such as China will develop the major (AI) automation manufacturing industry. The (AI) employers will need to employ many workers to manufacture any these different kinds of (AI) robots to satisfy China or overseas individual or business buyers needs. But, (AI) can also cause unemployment to the low skillful service workers. Such as if Japan some service businesses choose to buy any (AI) service robots to replace their service staffs to serve their clients. It is possible that the service staffs will be dismissed, due to (AI) robots can do such as their same service job duties to achieve better service performance. Thus, today, it is increasingly common for people to use robots in various situations at home and in retail stores, hotels and hospitals these service industries. Robots are classified into server types based on their functionality ( service and utility robots or those designed to communicate with humans) and appearance ( humanoid robots or mechanical robots). The type of robot, to which each country allocated particular importance in the advance of robotics, reflects the sense of values and preferences of its population. Thus, if the country has high population needs to use robots, then they will influence either more new jobs creation or more old job loss in the country's (AI) manufacturing or (AI) service industries both. For example, Japan respondents often associate the term " robot " with humanoid robots that can communicate with human and they have a high level of familiarity with robot. The US has the highest level of robot utilization at home and in retail stores with its people being the most enthusiastic about the future use of robots. Germany shows a strong tendency to consider robots for industrial purposes and its people feel strong effort to the presence of robots in their households.

In conclusion, to judge whether how (AI) will influence the country's employment to be better or worse. It will depend on the country home buyers (users) or business buyers (users) how to use (AI) for their daily needs. If the country , such as US retail stores need to use (AI) , it will have possible to reduce some or many retail service workers. Even, if the country , such as Japan has many home users need to use (AI) , it will not influence the employment market. Otherwise, it will raise (AI) salespeople numbers. Even, if the country, such as Germany and China will have many (AI) manufacturers, then it will create many (AI) manufacturing occupations for these (AI) manufactory workers.

Consequently, (AI) robots manufacturing and service needs will have positive or negative impact to any country's employment. It will depend on the (AI) service provision and service workers' job nature as well as the manufacturing workers of (AI) knowledge level to decide their employment chance in their country's employment market.

What does artificial intelligence(AI) mean?

- What (AI) function is?

Some scientists explain that artificial intelligence means which is an expert system, computer software that embodies a portion of the specialized knowledge of a human portion in a specific, narrow domain, owns decision making ability of human expert. The (AI)technology is based on the premise that what makes a person an expert is years of experience that enables who recognizes certain patterns in a problem as being similar to pattern. For example, in the future artificial intelligence system can be applied to control air traffic, design to computer configuration, medical diagnosis, instruction/training, speech/interpretation, monitoring to ( nuclear plant), planning to mission, factory scheduling, prediction weather, repairing telephone, automatic driving etc. different industries.

Artificial intelligence characteristics include: creative, adaptive , common sense, fact processing, quick replication, broad focus permanent and consistent skill. Otherwise, traditional computer expert system disadvantage includes perishable, unpredictable, slow reproduction, expensive, slow reproduction, slow processing lacks inspiration, needs instruction, narrow focus only machine knowledge. So, artificial intelligence is a branch of computer science devoted to creating computer to influence software and hardware to attempt to create human intelligence or human intelligent behavior. It is learning from experience, responds flexibility in situation that are, new or not anticipated.

Thus, (AI) can be learnt programmed knowledge to solve problems, using reasoning in solving problem, understanding and inferring facts and rules, recognizing the relative importance of different elements in a situation. In summary, artificial intelligence is concerned with two basic ideas mainly: The first idea, it involves studying the thought processes of humans to understand what intelligence is; the second idea, it deals with representing thought processes using companies to create artificially intelligent entities for testing the theories of intelligence.

- Can (AI) impact human job nature?

Human need concern this question: Will artificial intelligence (AI) reduce some human jobs in order to instead of replacing machines to do? Due to artificial intelligence is the ability of machines to do thing, that people would require intelligence. For example, artificial intelligence machine man driving( self-driver), it (AI) machine man driving research is an attempt to discover and describe aspects of human intelligence that can be simulated by driving machine functions. Alternatively, (AI) mathematical research may be another viewed as an attempt to develop a mathematical theory function to describe the abilities and actions of things ( natural or man-made) exhibiting intelligent behavior and server as a design of intelligent calculation machine function.

Why do humans need artificial intelligence machines to instead of traditional human service job? For example, can artificial intelligence machine man (self-driving) driver drive to replace human driver? I shall compare the differences between humans and computers : The characteristics of humans are good at recognizing various things, either seen before or not, recognizing the relationship patterns between things. Human thinking is common sense reasoning, combining all types of sensory input, acting appropriately in novel situations, learning new things and changing behavior patterns, making decisions , even when given incomplete information, working with noisy, incomplete information gathering behaviors . However, characteristics of computers are good at: The tasks humans do naturally are extremely difficult for a computer program as intelligent, which must be able to do the same kind of tack as humans do naturally.

Hence, (AI) is an combination of many different success and technologies: Linguistics - computational and socio, philosophy-logic, philosophy of mind and of language, electronical engineering -image and speech processing, pattern recognition, robotics, machine learning, neural networks, optimization scheduling, management information system and decision making. So, it is possible that (AI) can impact human job nature to instead of human working behavior in the future.

- How can human society job nature

to be changed to artificial intelligent society?

From the first intelligent perspective reason view point, artificial intelligence is making machines " intelligent" acting as humans expect people to act. Artificial intelligence has ability to distinguish computer responses from human responses, it owns knowledge to solve expert problem. From another research perspective reason view point, artificial intelligence is the study of how to make computers do things which, at the moment, people do better ( Rich & Knight, 1991, p.3).

(AI) researchers are native in a variety of domains, e.g. formal tasks (

mathematics, games), tasks ( perception, robotics, natural language, common sense reasoning), expert tasks ( financial analysis, medical diagnostics, engineering, scientific analysis and other areas).

From the second business perspective reason view point, (AI) is a set of many powerful tools, and methodologies for using those tools to solve business problems. From a programming perspective reason view point, (AI) includes the study of symbolic programming problem solving and search .

From the third human technological perspective reason view point, today's computer can do many well-defined tasks, for example, arithmetic operations, are much faster and more accurate than human beings. However, the computers' interaction with their environment is not very sophisticated yet. How can human test whether a computer has reached the general intelligence level of a human being? Can a computer convince a human interrogator that it is a human? But before thinking of such advanced kinds of machines, human will start developing our own extremely simple " intelligent" machines. So, it is possible that human society job nature will to be changed to artificial intelligent society when (AI) technology is developed to the mature stage in the future.

● Why does human need artificial intelligence machines?

One of major division in (AI) is between humans who think (AI) is the only serious way of finding out how we ( human) work and human who want companies to do very smart things, independently of how we ( human) work. This is the important distinction between cognitive scientists vs engineers. One of another major division in (AI) is between symbolic (AI), which represents information through symbols and their relationships. Specific Algorithms are used to process these symbols to solve problems or deduce new knowledge and connectionist. So ( AI) , which represents information in network. Biological processes underlying learning, task performance and problem solving are imitated from human mind behaviors. Thus, it is possible that artificial intelligence machines can do the better judgicious behavior to compare human.

● How does artificial intelligence influence future working changing in automation employment and productivity aspects?

In the automation changing influence aspect, as companies increasingly use robots on production lines or algorithms to optimize their logistics manage inventory, any carry out other core business functions. Technological advances are creating a new automation age in which ever-smarter and more flexible machines will be deployed on an ever larger scale in the marketplace. However, researching artificial intelligence with how influences human working nature. We need to answer these questions: How will automation transform the workplace? What will the implications for employment? And what is likely to be its impact both on productivity in the global economy and on employment?

Advances in robotics, artificial intelligence, and machine learning are growing in a new age of automation as machines match or outperform human performance in a range of work activities, including ones requiring cognitive capabilities. What factors are determined the changing in workplace adoption by artificial intelligence innovation? What advantages are automation? Automation of activities can be enabled businesses to improve performance by reducing errors and improving quality and speed, and achieving outcomes that go beyond human capabilities.

Some scientists indicated based on their scenario modeling. They estimated automation could raise producing growth globally by 0.8 to 1.4 percent annually. Almost, the activities people are paid almost $16 trillion in wages to do in global economy have the potential to be automated by adopting currently demonstrated technology. According to their analysis of more than 2,000 work activities across 800 occupations. When less than 5% of all occupations have of least 30% of activities that could be automated. They also indicated that technical economic and social factors will determine automation. Continued technical progress, for example, in areas such as natural language processing is a key factor beyond technical feasibility , the cost of technology, competition with labor including skills, and supply and demand dynamics, performance benefits including and beyond labor cost savings and social and regulatory acceptance will affect ( alter) the scope of automation.

Other some scientists also indicate U.S. country for example, the anticipate shift in the activities in labor force of a similar order of magnitude as the long term sight away from agriculture and decreases in manufacturing. Share of employment in the United States both which were achieved. So, those factors can influence why artificial intelligence technology needs. So, it is possible that future agriculture and manufacturing both industries will apply (AI) technology manufacturer-kind of job nature to raise productivity instead of farmers, fruit picking workers, farming transportation labours as well as factory manufacturing workers and supervisors etc. human-kind of job nature.

- Is artificial intelligence possible to replace labor ?

Not just intelligence, but also debating, if machines are capable of having a conscious minds. Artificial intelligence has those characteristics as below:

On functionalism aspect, artificial intelligence inputs mental states, sensory inputs, ( beliefs, desires being in pain feeling) and behavioral outputs. Since mental states are identified by a functional role, which are thoughts to be manifested in various systems. Even, perhaps computers which are physical devices with electronic substrate that inform computations on inputs to give outputs similar to brains which are artificial intelligence composed of part any intrinsic relationship to each other. Thus, artificial intelligence activities is not the whole itself, but into parts or on external influence on the parts.

On dualism aspect, artificial intelligence is a set of views about the relationship between mind are matter. On materialism aspect, it builds the only thing that exists is matter, including consciousness.

On biological naturalism aspect, it is similar a human brain than feels pains makes mental situation. So, artificial intelligence is similar biologist which might to be excited to human labor work. Hence, it seems artificial intelligence can change ( alter) or replace human labor work of nature in possible in the future.

- Can (AI) technology replace human labour nature of work?

On technological innovation reason view point, the history development of artificial intelligence studying the intelligence is one of most ancient scientific discipline. The history development of artificial intelligence what aims to achieve human use to sense, learn remember and think, logic probability, decision making and calculation develop from mathematics, instead of replacement human labor functions.

Artificial intelligence history development aim is the scientific analysis of skills in connection and practice with the appearance of computers from 1950 year beginning. The artificial intelligence (AI) can deal with the ultimate challenges. How can ( either biological or electronic) mind sense, understand and manipulate a world that is much simple and more complex than itself? And what if would human like to construct something with such capabilities?

The general-purpose software of the early period of (AI) were only able to solve simple tasks effectively and failed when which should be used in a wider range or an more difficult tasks. One of the sources of difficulty was that early software had very few or mix knowledge about the problems which handled, and activities successes by simply syntactic manipulation. Moreover, the other difficulty was that many problems that were tried to solve by the (AI) were untreatable.

The early (AI) software whether trying step sequences based on the basic facts about the problem that should be solved, experimented with different combinations till which found a solution. From the end the 1960 year, developing the so-called expert systems were emphasized. These systems had ( sue-based) knowledge base about the field which handled. Till to the beginning of the 1970 year, ( Prolog) the logical programming language was born, which was built in the computation realization of a version of the resolution calculus. ( Prolog) is a remarkably prevalent tool in developing expert systems ( on medical, judiciary and other scopes), but natural language parsers were implemented in this language. Then, in 1981 s, the Japanese announced the fifth generation computer system project, a 10 years plan to build an intelligent computer system that use the ( Prolog) language as a machine code. Nowadays, (AI) can be applied any industries, such as car manufacturing industry can use (AI) technological machine-men manufacture car, instead of replacing human labors in factory. Even, in the future, using (AI) machine-men drivers can drive any private cars or public transportation tools, instead of replacing human drivers, e.g. bus, train, tram, ferry etc. Also in the future, machine-men can replace housewives to serve families to do housekeeping clean job , e.g. cleaning toilets, bathrooms, kitchens, even cooking functions at home. So (AI) machine-man can reduce housewives works at home. Moreover, (AI) machine man can take care old people , when who are living at homes or elder care centers.

So, it seems artificial intelligence (AI) will be possible developed to manufacture a new generation machine-man to assist ( serve) families to do any simply cleaning or cooking jobs at homes. Moreover, the overall demand of ( AI) general social needs will also rise, such as security, driving transportation tools, restaurant cleaning, elder centers care service etc. So, it seems that individual or families or social needs of (AI) will be increase in the future. Thus, it will influence macro economy growth (GDP) if there are large house family consumer group and hotel or bus or taxis or ferry etc. different business consumer group demand any artificial intelligence machine numbers increasing. Then, the artificial intelligence products and material manufacturers must need to buy many artificaial intelligence materials to produce any kinds of artificial intelligence machines to prepare to satisfy consumer individual needs. Consequently, macro economy will grow to the owned artificial intelligence development countries, e.g. US, China, UK.

- Why can artificial intelligence satisfy human needs?

First, On machine-man satisfactory demand aspect view point, it makes computers that think, it is the automation of activities. We associate with human thinking: like decision making, learning. It is the act of creating machine that perform function that require intelligence when performed by people. It is the study of mental faculties through the use of computational models. It is the study of computations that make it possible to perceive, reason and act. It is a branch of computer science that is concerned with the automation of intelligent behavior. It is anything in computing service that human don't yet know how to do property.

Second, on thought aspect artificial intelligence means systems thank think like humans, systems that think rationally.

Third, on behavioral aspect, artificial intelligence systems that act like human and that systems act rationally. However, the basic objective of (AI) is to represent human's thought processes in computation . These machines are supposed to exhibit behavior that. It is performed by a human being, would be considered intelligent. However, some authors feel (AI) has disadvantages, such as it is not creative, it is excited in the use of sensory devices, it can't make use of a very wide context of experiences and it does not use common sense.

For speech recognition and understanding function needs example, (AI) can be applied in speech recognition and understanding function, which (AI) speech or voice recognition is a data input method. For example, the computer recognizes and understands one ( or a few) word commands. Speech understanding on the other hand is the computer's ability to understanding a spoken language. That is , the computer understands the meaning of sentences, an paragraphs through (AI).

So, (AI) can be attempted to learn human language how to speak. It is similar to translate human language skill, instead of actual human speaking skill. Also, (AI) can assist handicap learning or language student how to listen different languages by machine-man sounds from computers more accurately. So, it seems that it (AI) can replace human language teachers speaking function and can change teaching language nature of job in language speaking and listening education industry.

- Is artificial intelligence one good choice for human future technological benefit?

Nowadays, new technology development is popular. However, artificial intelligence is one kind of new technology choice among different technologies innovation. So it brings this question: Is artificial intelligence technology value to invest? To answer this question. I shall indicate some other new technology developments to compare (AI) technology development to judge which has urgent needs to achieve human expectation nowadays.

For example, why is green peace interested in new technologies? New technologies features prominently in our ongoing campaigns against genetic modified crops and number power. However, which are also an integral part of our solutions to environmental challenges, including renewable energy technologies, such as solar, wind and wave ( water) power energy as well as waste treatment technologies, such as mechanical, biological treatment. It seems humans need concern how to apply (AI) technology to solve environment pollution challenges in our future. So, environment protective, agriculture, natural energy technology will be popular demand to attempt to apply (AI) technology to solve their challenges or apply (AI) to assist to develop their industry.

CHAPTER TWO

# AI human mind learning second stage

● How can artificial intelligence technology influence economy?

Advances in artificial intelligence (AI) technology and related fields have opened up new markets and new opportunities progress in critical areas, such as health, education, energy, economic development, social welfare and the environment pollution.

(AI) automation will continue to create wealth and expand the global economy development in the future. However, when many will benefits that growth won't be costless and will be accompanied by changes in the skills, that workers need to increase productivity in the economy and structural changes in the economy. So, in the skills that workers need to succeed in the economy and structural changes.

I shall indicate why aggressive policy action will be needed to help Americans who are disadvantaged by these changes , due to (AI) technology is caused. For automation industry change example, artificial intelligence (AI) capabilities will enable automation of some tasks that have long required human labor. These artificial intelligence technology introduction can increase new opportunities for individuals. The economy and society, but (AI) has also the potential to disrupt be current livelihoods of many Americans. However, (AI) leads to unemployment and increase in inequality over the long run depends not only on the (AI) technology itself, but also on the institutions and policies that are changed. Thus, it is possible that (AI) technology will raise some countries unemployment number if the employer apply (AI) technology workers to work instead of human labor in their factories, but it can also raise productivities for these employers.

● Can (AI) influence global economy growth?

Technological progress is main driver of growth of GDP per capita, allowing output to increase faster than labor and capital . However, technology can increase productivity, but also decrease the number of labor hours needed to create a unit of output. So (AI) causes unequal to labor wage decreases, even reduces the number of labor to manufacture, e.g. artificial intelligence technology of automation car manufacturing industry; clothing manufacturing industry; plane manufacturing etc. high technology of artificial intelligence manufacturing method. But (AI) should be potential environment benefit, although it raises unemployment ratio. Moreover, it can rise production , due to many skilled craft were replaced by the combination of machines and lower-skilled labor. The result of (AI) technology introduction , it causes output per hour risen when inequality declined, driving up average living standards, but the labor of some high-skill workers was no longer as valuable in the market. Otherwise, if (AI) technology is continue developed to be success. Some routine intensive occupations will be loss, which focused on predictable, e.g. easily programmable tasks, such as switchboard operators, filing clerks, travel agents, and assembly line workers would be particularly replaced by new (AI) technology. However, at the same time, (AI) technology development will bring these benefits: improvement in education ( training (AI) technology scientists) , due to (AI) manufacturing technology needs are raising to businesses and institutional changes, such as the reduction in unionization and raising in the minimum wage to the (AI) manufacturing technology skilled labor in factories.

Because (AI) technology is not a single technology, but rather a collection of technologies that are applied to specific tasks, the effects of (AI) will be felt unevenly though the economy. It will bring some tasks will be most easily automated than others , and some jobs will be affected more than others, both negatively and positively. Finally, new jobs are likely to be directly created in areas , such as the development and supervision of (AI) as well as indirectly created in a range areas though out the economy as higher incomes lead to expanded demand.

However, if (AI) technology could dominate global labor markets. If labor productivity increases, do not influence into wage increases, then the large economic gains brought about by (AI) technology could be increased wealth inequality, due to employers can reduce production cost, but workers ( labors) wages will not be increased, even will be decreased. Hence, it seems the (AI) technology will bring disadvantages to labor market to cause unemployment or reduce wages in possible, although it can reduce employer individual salary (wage) expenditure and it can raise productivity.

● How can artificial intelligence impact global economy growth?

Artificial intelligence (AI) technology is a branch of computer science that aims to create intelligent machines that work and react like humans. So, (AI) is a technology that appears to impact ( influence) human preference by learning, understanding complex contents, enhancing humans in executing both routine and non-routine tasks. In the future, (AI) technology that can be virtual personal assistant, as well as it may exist, such as robots with human-like processing capabilities.

How can (AI) technology impact global economy growth over the next 10 years? During this time period, (AI) technology is predicted to have wide-ranging applications including: Machine learning that automates analytical model building by using algorithms that allow machines to operate without human assistance.

In global education aspect, potential applications include predicting cause-and-effect relationships from biological data, identifying new drugs, self-driving cars, and protecting against fraud, improved natural language processing that allows computers to continue to better analysis, understand and generate language to interface with humans using natural human languages. For example, transcribing notes dictated by physicians, automatically drafting articles and translating text and speech. So (AI) technology can be applied to education aspect to improve humans' knowledge level.

In visual art aspect, (AI) machine vision that allows computers to identify objects, scenes and activities in images. Current applications of (AI) machine vision include providing objective descriptions for the blind seeing( visual) needs.

We except the economic effects of (AI) technology to include both direct GDP growth from sectors that develop or manufacture. (AI) technology and indirect GDP growth through increased productivity in existing sectors that employ some form of (AI). If (AI) technology is an increasingly critical component of more products, it will become an integral part of many people's lives. Thus, (AI)'s ability to influence economic activity, rather than the economic or development status of the region. (AI) has the potential to impact income classes and to bring significant gains to both developed and developing countries. For example, (AI) has the potential to optimize good production around the world by analyzing agricultural regions and identifying what is necessary to improve crop yields.

In estimating the future economic effects by (AI) technology innovation, it is important to note that it is challenging to accurately predict which applications of (AI) will ultimately be commercially successful. In micro level economic influence, we need to apply methodologies to estimate the economic effects of investment in firms developing (AI) technology since investment levels in a technology are a telling sign of the future potential of that (AI) technology.

● How can (AI) influence GDP of high income countries in the next ten years?

How (AI)'s development may affect the global economy over the next ten years. In fact, (AI) technology has the potential to affect business across the global in a wide range of industries in ways only a number of technologies have done in the parts. For example, (AI) technology's expected to be a useful tool for enhancing human capabilities and in some instances replacing functions, such as driving a car, adoption of broadband internet, mobile telephone, industrial robotic automation have served to enhance human capabilities.

However, significant public debate has focused on projections of (AI) technology's effect on the labor force. However, large companies prefer to invest in (AI) technological industry. For example, face book's (AI) research lab., google machine intelligence lab. and micro soft machine learning and artificial intelligence research division are all making advances in (AI) technology and investing in the industry's top talent. Additionally, between 2010 year and 2015 year, nearly $5 billion in venture capital funding invested in firms across the global developing and employing (AI) technology ( Facebook (AI) Research).

● How can artificial intelligence impact on workplace?

Modern information technologies and the labor economy growth of machines is powered by artificial intelligence have already strongly influenced the world of work in the 21 ST century. Computers, algorithms and software simplify every tasks and it is impossible to image how most of our life could be managed without them. How can be the information economy characterized by exponential growth replaces the most production industry based on economy of scales? What will the future world of work look like and how long will it take to get? Will the future world of work be a world where humans spend less time earning their livelihood? Alternatively, are mass unemployment, mass poverty and social distortions also possible scenario for the future, where robots, artificial intelligence systems play an increasingly central role? These questions concern how artificial intelligence further development . Can influence labor economy growth on workplace ? When the labor market has widespread impact on intelligence property, information technology, product liability, competition and labor and employment laws.

How (AI) technology impacts on labor workplace.

The future influence any organizations how labor economies use of (AI) can be analyzed, such as deep machine learning is based on a set of model high level data. Unlike human workers, the machines are connected the whole time in workplace. If one machine makes a mistake, all autonomous systems will keep this in mind and will avoid the same mistake the next time.

Over the long run intelligent machines will win against every human expert. Production robots have been replacing employees because of the (AI) technology. They work more precisely than humans and cost loss. Creative solutions like 3D printers and the self learning ability of these production robots will replace human workers, the automatic data recording and data processing, traditional back office activities are no longer in demand. Autonomous software will collect necessary information and will send it to the employee who needs it. Additionally, dematerialization leads to the phenomenon that traditional physical products are becoming software. For example, CD or DVDs are being replaced by streaming services. The replacement of traditional event ticket, e-travel ticket service products or hard cash will be the next step, due to the possibility of payment by smartphone. So, (AI) technology will impact human's daily life consumption behaviors in the future. For another example, transportation tools, such as boats and ferries and private vehicles will use sensors and navigating without human input. Taxi and truck drivers will become obsolete, the stock store applies to stock managers and postal carriers of the delivery is distributed by (AI) machine delivery method,

2.1`What is the relationship between (AI) and (CRM)?

● Can (AI) technology impact on customer relationship management (CRM) ?

Nowadays , (AI) is a technology almost as old as the computer industry itself, it is similar with the advent of personal assistants function to businesses and personal promotion channel, such as ( Amazon's Alexa, Apple's Siri, Google's Assistant) image recognition ( face book),

personalized recommendations ( Netflix , Amazon). Those innovations have been driven by a increase in processing power, lower cost hardware, and the exploding creation and availability of data. It seems, (AI) technology can impact global customer service management method.
How to forecast economic impact modeling to (AI) will affect global economy? Can human forecast business revenue growth and job creation ( or destruction) based on (AI) applied to customer relationship management (CRM) activities? In addition to the economic impact on (AI) or (CRM) which can include an estimate of the economic impact attributable to sales forces customer base. What can economic benefits be brought to (CRM) from (AI) technology?
Artificial intelligence(AI) comprises a set of technologies that use natural language processing, machine learning, knowledge graphs, and other tools to answer questions, discover insights and provide recommendations. Computer systems can use (AI) hypothesize and formulate possible answers based on available evidence can be trained through the ingestion of vast amounts of content, and automatically adapt and learn from (AI) self mistakes and failures.
So, any business organizations (customer service departments) can provide efficient and effective customer relationship management of excellent customer service quality if which applied (AI) technology system. The different type of (AI) systems include: (AI) system platforms, machine learning (AI) based data preparation and enrichment tools, machine vision/ image recognition, voice speech recognition, text analysis and natural language processing, bots , e.g. face book website and virtual digital assistance solutions, social media pattern analysis , sentiment analysis, advanced numerical analysis (e.g. IOT streaming , machine logs), supporting technologies, knowledge base dialog management, Q&A processing etc. different (AI) technology system customer relationship management (CRM) tools.
(AI) (CRM) of activity can include these categories, such as: corporate marketing, marketing operation, field marketing, customer support, digital commerce, customer analytics, customer influenced product or service design, product or service pricing, finance information, presentation, customer billing, inventory , logistics and fulfilment support, partner management etc. different CRM tools.
(AI) technology of CRM has been carrying on plan different stages to achieve CRM personal assistant tool for businesses. The stages are such as,

in the beginning stage of (AI) projects in place, implement now, pilot phase next year in the final stage of (AI) customer relationship management tools are foreseeable future. So, this CRM technology has been improved to plan in different stages every year to prepare to achieve full capacity of CRM service quality for businesses to use in the future.

Hence, how to develop an estimate prediction of the economic impact (AI) technologies could have CRM activities, which depends on gathering macroeconomic information on business revenue and the basic marketing of business revenue and the basic markup of business expenses by major functions ( customer support, marketing and sales , production etc.)

An economic impact model that can gather data together and forecast the results how (AI) artificial intelligence technology brings (CRM) customer relationship management benefits to businesses, e.g. surveys investigation includes IT spending by sample countries, GDP and population estimates and forecasts, revenue per employee and ratios of IT spend to GDP. Surveys ( questionnaire questions) of forecast results are influenced by (AI) impact can include: results are projected from surveys and rely on estimates are made by respondents on the expected financial improvements in categories of (AI) –assisted customer relationship management activities. The forecast assumes that these estimates are correct; financial estimates are based on estimates of "first year" improvement from full (AI) implementation; forecasts are from planning to implement any artificial intelligence of customer relationship management (CRM) projects, the improvement forecast is of categories of activity , e.g. corporate marketing , digital commerce, and customer analytics. They are not estimates of ROI for the (AI) software. They rely on conservative estimates to which each of these entities might affect company revenue, expenses or productivity. They also rely on estimates of the penetration of software in customer relationship management activities . Net new jobs created are based on the ratio of new revenue to jobs required to support that revenue . They can assume that 50% of the net new revenue will support increases in labor and the rest will go for capital and other operating expenses that may replace jobs lost to automation.

In the future, some of the ways in micro economic benefits to any organizations. (AI) technology is expected to impact CRM activities include: Spending up sales cycles, improving lead generation and qualification solving customer support problems faster ( raising service quality), helping companies improve brand campaigns and recognition, lowering costs of support calls when increasing resolution rates, lowering the cost of recruiting employees and partners, increasing revenue from optimized product marketing, optimizing price, distribution logistics and preventing loss through fraud detection. So, micro economic benefits view point, it seems that (AI) CRM technology can raise any companies economic benefits for care term.

Artificial intelligence enables machines or the in-build software to behave like human beings which allows these decisions and act. The advent of (AI) is leading , talking, making decisions and act. The advent of (AI) is leading to new technologies advances and transforming the economic and employment opportunities for humans in a positive way. (AI) related technologies can facilitate our live. For example, industrial robotics, robotic medical assistants, smart games, financial forecasting software, big data analysis, algorithms in health and bioinformatics, pilotless cargo places, drone ambulances and general purpose and workplace robots and others. (Disruptors technologies: Advances that will transform life, business and the global economy).

Artificial intelligence also known as computational intelligence is defined as " the human –like intelligence exhibited by machines or software. It is theorized that intelligence of humans can be described and intelligence machines or software can simulate it. These machines software can be reasonable , learn, perceive and process information, like human mind and thus facilitate human life. They can think and act for us. So, artificial intelligence is an interdisciplinary field of study including computer science, neuroscience, psychology, linguistics and philosophy.

However, (AI) research and developments have economically impacted many industries, such as robotics, telecommunications, computer applications , health, finance, heavy manufacturing, transportation, aviation, e-service and e-commerce, military , music and movie, toys and games entertainment etc. industries.

In fact, many ideas, systems and technologies have been developing in the world of (AI) technology. However, which are net called or considered (AI) products, rather which are mentioned with their specific names, such as smart graphics, machine learning, e-commerce etc. ( i.e. this is called (AI) effect).

- How can (AI) technology influence digital economy?

Nowadays, (AI) related industrial applications will replace most human power in fields, including call centers, customer services and air cargo transportation. (AI) technologies also help weather forecasting based on repeated rainfall pattern ( data) recognition, through robotics ( i.e. floor cleaning, moving lawns etc.) transporting people and products with unmanned vehicles, sending space unmanned smart shuttles, developing robotic arms, predicting market values in stock exchanges by internet, making homes safer, helping elderly and disabled using robotic servants etc.

Among the (AI) related technologies , there are a few that significance for the impact on society and especially on digital economy . (AI) is particularly influential in machine learning. Such as robotics, transportation, finance, health and bioinformatics, e-commerce , e-games, big online data gathering and internet-of-things. For example, machine e-learning is based in bioinformatics and robots that can learn new skills for better caregiving in healthcare. What is machine e-learning? Machines can e-learn from e-data gathering, coming up generalizations and making decisions to act in certain ways from internet.

There are important applications , such as e-machine perception, electronic online natural language learning processing, online search engines, online bioinformatics, online brain –computer interface, online game playing, online robot locomotion, online advertising, online computations finances, online health monitoring, online DNA classification and decision making, online in chemistry –cheminformatics . So, online machine learning can positively impact productivity and it can enhance information and analytical system from (AI) online channel.

What is robotics? Robotics is one of the most strongly influenced fields in (AI). For example, heavy manufacturing industries, robots and used and man power is replaced for effectiveness, precision, and accuracy, especially in respective or dangerous tasks, including welding, assembling , picking and placing .

So, robots can acquire new skills or adapt the changing dynamic environment. Also, artificial intelligence can be applied in developing transportation. For example, automated vehicles, driver assistance systems , safety systems, collision avoidance systems and public transportation. Moreover, (AI) technology has proven to produce some of the best tools to predict stock market fluctuations from internet data gathering method. It's predictions are based on ever-evolving predictions algorithms and systems learn new models and make connections between historical data and new data to measure stock market trading more accurate from internet data

gathering channel.
In health field, especially in health data processing , analysis, decision making support and medical diagnosis. So, online data can show which patients will need what treatment and what alternative drugs could be used more accurate from (AI) online data gathering method. Bioinformatics is an interdisciplinary field combining statistics, (AI) online technology can help in discovering data patterns and modeling through the application of machine learning, artificial neural networks and genetic algorithms. For example, further (AI) technology development of human genome project of online data sequences.
Online shopping can be facilitated by virtual assistants developed through (AI) technology and these assistants can offer the best advice. (AI) online purchase coming after every product image recommendations and personalization bring important revenue to shopping online sites, like Amazon . Smart computer graphics and games, artificial intelligence is useful in smarter computer, graphics, scene modeling , scene rendering processes in order to create, for example, effective human –robot interactions , online machine learning, online strategic games techniques etc. online computer related (AI) software.
So, online big data analysis and big data does have a critical need in the world of online intelligence machines and software in our future. In other words, (AI) offers online technology to enable online big data analysis to provide industrial organizations with valuable information for effective decision making in short time. For example, what IBM's Watson achieved: this machine used 200 million of structured and unstructured content with a special technology of hypothesis generation, massive evidence gathering, analysis and scoring from internet channel.
Finally, (AI) online technology another related internet invention ( internet of things) (IOT) is the network of machines or objects connected through internet. These connected objects can sense their internal and external environment, communicate with each other, can send critical data and finally can make decisions to act or correct their environment from (AI) online technology. For example, factories can monitor and automatically change production processes, hospitals can monitor and regulate the health conditions of their patients , schools can collect data from facilities and cars can send data to car makers from (AI) online technology.
Partner predicts that (IOT) market will create about trillion amount value by 2020 year. Although machines collect big data from their environment,

whether which gain an insight or learn from these online data largely depends on the (AI) online machine learning principals and (AI) online technology. In 2013, Mckinsey estimated that disruptive technologies closely related with potential economic impact in 2025 year between $7.1 to $13.1 trillion amount ( automation of knowledge work, advanced robotics, autonomous or near-autonomous vehicles).

What is the relationship between
(AI) and global digital economy development ?

● Could work activities in China be automated
making in the nation with the world's largest automation potential?
Can (AI) technology influence China economy? Could China workers be affected and jobs made up of routine work activities and predictable? Will programmable tasks be particularly impact to China employment market ? When impact on labor market is likely to be gradual at the aggregate level, it can be sudden and dramatic at the level of specific work activities, rending some job obsolete fairly. Overall (AI) technology will raise digital skills when reducing demand for medium incomer inequality for China workers. It seems (AI) technology's effect on productivity could be crucial to China's future economic growth as the population ages are increasing.
In China, some biggest technological companies driving significant investments in research and development. Moreover, China is one of the leading global (AI) technology development county. However, China will need to focus on building its innovation capacity. For example, United States and United Kingdom are currently producing more influential (AI) technological research. However, if China planed to achieve (AI) technology success, it's traditional industries will need to develop technical know-how –to and overcoming implementation costs prepare to develop (AI) . When (AI) technology is introduced into China society, China government needs to raise concerning ethical, legal, technological security etc. business questions. Also, surrounding issues include privacy, discrimination, legal liability and regulation. It aims to encourage overseas investors to choose to invest (AI) technological industry to raise GDP growth and manufacturing industries income growth for long term in China. If China encouraged overseas (AI) technology investment in its country. It is possible to influence China employment market to be changed. Because (AI) technology will impact to influence China people daily life. Due to (AI) technology is introduced to China society, many rich people will prefer

to spend to buy any high (AI) technological products for entertainment or learning or machine man driving etc. daily necessity activities. Then it will raise GDP growth and will raise (AI) manufacturers or related-(AI) technological manufacturers profit. It is beneficial to China because it can become one high knowledgeable and (AI) technological economical society. But it will bring bad influences to raise unemployment chance for the low skillful labor. In labor economy aspect influence , how (AI) technology can influence China low skillful labor unemployment ratio raising. The raising low skill labor unemployment reason is because China low skillful human labors are argued or are replaced by (AI) technology creating new challenges to introduce to influence China society of simply human manufacturing job nature to be changed to be high (AI) technology manufacturing job nature in any China factories. Moreover, when (AI) technology introduction to China, it will cause other related social challenges in China. The varied (AI) related challenges, including the difficulty of creating safe and reliable hardware for sensing and affecting ( transportation and education), the challenges of gaining public trust, a low resource comities and public safety and security, the challenges of overcoming fears or marginalizing humans in China employment and workplace and the risk of diminishing interpersonal trust because the low skillful labors won't believe any China employers will give chance to employ them , due to (AI) technology will replace their skills and man manufacturing of productivity is much less to compare to (AI) technology manufacturing method.

- How does ( AI) technology influence
the future of employment change?

Are future nature of jobs changed to computerization from (AI) technology? Where are the probability of computing occupations from (AI) technology influence? What is expected impacts of future computing on labor market from (AI) technology influence? John Maynard Keynes's frequently cited prediction of widespread technological unemployment " du to our discovery of means of economic the use of labor outrunning the pace of which we can find new used of labor" ( Keynes, 1933, p.3).

In the future, (AI) technology will impact some nature of occupations to change computing. This chance will also influence some countries' economic change. For example, some factory human labors hand routine manufacturing tasks will be changed to computerization of routine manufacturing tasks by (AI) technological machine men hand manufacturing method. it will cause a structured shift in the labor market, with workers reallocating their labor supply from middle-income manufacturing to low-income service occupations.

Arguably, this is because the manual tasks of service occupations are less computerization, as who require a higher degree of flexibility and physical adaptability. So, (AI) technology will influence the human hand labor skillful occupation nature of task cheaper , such as vehicle manufacturing , ship manufacturing, computer manufacturing, steel manufacturing, television, radio etc. home electronic products of heavy machine industry change. Due to (AI) technology machine man will be proper to be used to manufacturing these electronic products when the (AI) technology innovation can develop to the mature stage. Then, any countries manufacturers will choose to use (AI) technology machine man, instead of human hand production.

Supposing the future prices of computing are fallen, seriously, problem solving skills are becoming relatively productive, explaining the substantial employment growth in manufacturing occupations, involving cognitive tasks where skilled labor has a comparative advantage, as well as the increase education needs for (AI) technology computing of machine man subject study.

Prediction of education needs for (AI) technology student numbers will increase, due to manufacturing industry needs many (AI) technology students in future employment market. Another (AI) technology influence if the future (AI) technological innovation, e.g. machine man manufacturing or machine man service industries will both increase demand, then with more sophistic software technologies will be disrupted labor markets by marketing workers redundant.

For publishing industry, what is striking about the case in paper book publishing industry will be unpopular? Due to the electronic book publishing industry will be popular, e.g. Amazon publish . (AI) technology can influence paper book manufacturing method which is replaced by machine man electronic book manufacturing method as well as it will cause the computerization is no longer confined to routine manufacturing tasks. Due to (AI) machine man manufacturing technology will be proper to be used to manufacture any products in short time efficiently and effectively , e.g. electronic book products. In the future, if it is fact to occur this case, such as ( AI) technological machine man manufacturing method will be adopted ( applied) to manufacture electronic books or any products in possible. (AI) technology will cause many manufacturing workers are unemployed. It is beneficial to employers, who can reduce to spend much wages expenditure to employ manufacturing workers, but it will cause many manufacturing workers loss jobs and reduce income to support whose families lives. It will cause social challenges, e.g. increasing stealing crimes if the manufacturing workers had not other skills to find other jobs to do easily. So, manufacturers need to concern over technological unemployment which will be hardly future phenomenon if who decided to dismiss all manufacturing workers, due to (AI) technology machine men replace to them.

If ( AI) technology can be innovated to produce any kinds of machine man to serve any service or manufacturing industries successfully. Then, it will bring these questions: Can future that workers be influenced to be automation employment and productivity by (AI) technology influence? Does it impact to influence the (AI) technology countries' productivity and growth and natural resources development and labor markets and evolution of global financial markets and economic impact of technology and innovation and urbanization etc. issues? How will automation transform the workplace? What will be the implication for employment? What is likely to be its impact both on productivity in the global economy and on employment?

In fact, automatic of activities can enable businesses to improve performance by reducing errors chance and improving quality and speed, and same cases achieving outcomes that go beyond human capabilities. Some economists indicate (AI) technology would give a needed boost to economic growth and prosperity have of the working age population in many countries. Based on the scenario modeling, they estimate automation could raise productivity growth globally by 0.8 to 1.4 % annually. They also indicated that almost half the activities people are almost $1.6 trillion in wages to do in the global economy have the potential to be automated adapting current demonstrates technology, according to their analysis of more than 2,000 work activities across 800 occupations. When less than 5% of all occupations can be automated entirely using demonstrated technology, about 60% of all occupations have at least 30% of worker made activities, that would be automated. More occupation will change to be automated. They also indicated for business performance benefits of automation are relatively clear, but the issues are more complicated by policy making to attract foreign investors. Beyond technical feasibility, the cost of technology, competition labor will include skills and supply and demand dynamics, performance benefits and beyond labor cost savings and social and regulatory acceptance will affect the automation. Their predictions suggest that half of today work activities could be automated by 2055 year, but this could happen 10 to 20 years earlier or latter depending on the various factors in addition to their wider economic condition.

Some scientists suggest (AI) technology is finally starting to deliver real-life business benefits. Computer power is growing significantly , algorithms are becoming more sophisticated and perhaps most important of all, the world is generating vast quantities of the fuel that powers (AI) technology data billions of gigabytes of it every day. Also, online firms are digital natives, such as Google online search service company is investing on (AI) technology. For new though most of the news if coming from the suppliers of (AI) technologies. And many new users are only in the experimental phase. Few products are on the market or are likely to arrive these soon to drive immediate and widespread adoption. As a result, analysts believe (AI) technology's potential will give true economic benefit in the future. (AI) industry will introduce to suppliers and users to raise economic potential of (AI) technology.

In the future, (AI) technology systems can solve business problems. Some scientists categorized those into five technology systems that are key areas of (AI) technology development: robotics and autonomous vehicles, computer vision language virtual agents and machine learning , which is based on algorithms that learn from data without replying on rules-based programming in order to draw conclusions or direct an action.

Such as computer vision and language includes natural language processing, analytics, speech recognition technology, some are about learning from information, such as about machine learning and others are related to acting on information, such as robotics, autonomous vehicles and virtual agents, which are computer programs that can converse with humans. Machine learning and a subfield called deep learning are artificial intelligence applications.

- Can artificial intelligence impact
global economy growth?

Artificial intelligence (AI) is a term first defined in 1956 year. It is a branch of computer science that aims to create intelligent machines that work and react like humans. In contrast today, 60 years later, (AI) is characterized by a number of applications, including computers playing games against humans and understanding human languages, virtual personal assistants, and robotics which involve computers seeing , hearing and reacting to sensory stimuli. In the future, technologists predict for (AI) technology ranging from (AI) being used as a tool to aid relatively simple processes for robots with human like mental capabilities, who expect (AI) technology can emulate human performance by learning, coming to mind its own conclusions, understanding complex content, engaging in dialog with people, enhancing human cognitive performance or replacing humans in executing both routine and non-routine tasks. In existing industry, (AI) technology is used , such as targeted advertising and virtual used personal assistant as well as the (AI) technology that my exist in the future, such as robots with human vehicle processing capabilities.

The range of (AI) technology's progress in the future will determine the economic impact future of (AI) technology on the global economy with more limited advances and applications ( i.e. weak (AI) only) corresponding to more limited economic impacts and more substantial progress, i.e. strong (AI) technology is corresponding to more significant economic impact.

(AI) technology learning that automates analytical model, including predicting cause-and-effect relationship from biological data, identifying new drugs, self-driving cars and protecting against fraud etc. functions. Also (AI) learning can improve natural language processing that allows computers to continue to better analyze, understand and generate language to interface with human using the natural human language, virtual personal assistant, helps users by providing scheduling appointment, reminds organizing personal finance and finding providers of various services, machine vision allows (AI) machine man to identify object, scenes and activities in detect pedestrians and bicyclists.

We expect the economic effects of (AI) technology to include both direct GDP growth from sectors that develop or manufacture (AI) technology and indirect GDP growth through increased productivity in existing sectors that employ some from of (AI) technology. If (AI) producing sectors could grow, then it could lead to increase revenues and employment of (AI) technological professionals within these existing firms as well as the potential creation of entirely new economic activities to any countries' societies productivity improvement in existing sectors could be realized through faster and move efficient processes and decision making as well as increased (AI) technological knowledge and access to information available in societies easily.

In the future, if (AI) technology is an increasingly critical component of more products, it will become an integral part of necessary products of many people's lives. The extent of (AI)'s economy effort is also likely to vary from region to region, thought variation may be more dependent on the predominate economic activity of a region and the (AI) ability can influence economic activity, rather then the economic or developmental status of the regions. (AI) technology can move accessibility and can use source development to do international business between one country and another country.

So (AI) technology has the potential to give benefits to different income chooses and to bring significant gains to both developed and developing countries. For agricultural technology, (AI) has the potential to optimize food production around the world by analyzing agricultural regions and identifying what is necessary to improve crop yield. In total, (AI) technology gives greater economic impact to any countries agricultural regions if which implemented (AI) technology to grow crop , fruit etc. food production in the farms.

Investment in (AI) technology is such as capital investment to any countries‘ public or private enterprises. So, it will have large economic impact to the future . If the (AI) technology is reasonable invested to the different needs aspect by the public or private enterprises in the country. Then, it will have good economic impact to the country in the future. However, when (AI) technology is likely to affect both the productivity and employment components of economic growth in many sectors. Significant public debate has focused on projections of (AI)'s effect on the labor force. However, for instance, some researchers have argued that the rise of (AI) technology and automation will led to significant unemployment as capital is substituted for the low skillful labor. So, they point to the concern that the increasing sophistication of (AI) technology may balance skilled and semi-skilled workers and the reduce the size of the middle class. However, this is not a new argument, due to (AI) technology negatively affecting the labor force and leading to mass unemployment. Because the (AI) technology is the substitution of machinery for human labor. Although, employment in certain industries, has been reduced in the past due to technological advancement. For long term, the labor market has adapted to the introduction of new technology, giving rise to new jobs in new areas. (AI) technology may also be accomplished without a reduction to total employment in the long-term to some Asia countries, such as Hong Kong and Japan. Because Hong Kong and Japan many low skilled labor, e.g. security, cleaner who complaint that employers need them to work long time hours. ( abnormal working hours) e.g. one day 12 to 15 working hour per day. Hence, if (AI) machine means invention technology success. Security or cleaning job can be worked from (AI) machine man in some hours every day in order to reduce the long time working hours cleaners or security workers, e.g. one (AI) machine man works 4 hours for cleaning or security job, one day as well as another cleaner or security labor only needs to work 8 hours one day. So total security or cleaning employers can employ 12 hours machine cleaners or security workers and human cleaners or security workers in one day. For long term benefit, Hong Kong or Japan every security or cleaning worker does not need to work 12 hours minimum working hours one day. They won't feel tried and bore and without private with whose families, so who will accept to do these cleaning or security jobs, even they can raise work efficient and performance when who feel happy and health.

So, (AI) technology of machine man invention can raise low skillful labor efficiency and it can help them to avoid abnormal working hours demand in some busy work life countries, such as Hong Kong and Japan. Before, one Japan female labor feel unhappy to work, due to who often needs to work abnormal working hours for her employer and who has less sleeping and without any private time to enjoy her life with her families every day. So this abnormal working hours factor causes her to do commit suicide behavior, then she is die unlucky. So (AI) technology of machine man invention ought avoid abnormal working hours demand for employer in any countries in the future.

The most important occurrence to any employers, some researchers had attempted to do one experiment to find that private research and development , venture capital and public research and development investment all have strong net effect or economic growth with venture capital funding further having the strongest such effect from (AI) technology. The researchers hypothesize the venture capital investment contributes to economic growth through (AI) technology innovation and by the capacity of an economy to use existing (AI) technology knowledge to increase productivity. They predict the impacts of venture capital, business-research and development and public research and development can raise multi factor productivity from (AI) technology introduction.

Can (AI) technology influence the economic development to developing countries? The developing regions of the world contain most of natural resources. If one day, (AI) technology has invent one kind of machine man which can assist any gas or oil workers to seek any new oil/gas natural resource locations easily. I believe that (AI) technology can help these natural resource exploitation countries will gain economic benefit more easily. So, (AI) driven technology can be used to change to create any new opportunities to address poor management or resources and improve human well being, such as Africa Latin America and India can use (AI) technology machine man to seek any oil/gas natural resource countries exploitation activities to attempt to gain much economic benefits.

● Why will (AI) technology grow economic
development ?

Nowadays, increases in capital and labor are no longer driving the levels of economic growth, such as (AI) technology. The ability of increase in capital investment and in labor of traditional drivers of production, have no longer to be enjoyed in most developed economies ,e.g. developed country, US, UK . However, artificial intelligence has the potential to overcome the physical limitation of capital and labor to avoid missing out on this opportunity. So, policy makers and business leaders must prepare for and work toward a future with artificial intelligence. They must do with the idea that (AI) is another simply method to enhance productivity method . Rather they must see (AI) as the tool that can transform thinking about how growth is created.

Economists have always thought of new technologies are as driving growth their ability to enhancing. It can replace labor and capital factor of production. So, it brings this question: What is the factor of production (AI) technology characteristics. They key factor is to see (AI) technology as a capital-labor .

(AI) can replicate labor activities at much greater scale and speed, and to even perform some tasks began the capabilities of human. For example, by using virtual assistants , 1000 legal documents can be reviewed in a matter of days instead of taking three people six moths to complete. Some (AI) technology may be one kind of factor of production in the future. For another example, people will work in workplace digitalization environment. So, in the future, working environment and information management are automated. Such as Konica camera sale company will use workplace digitalization. So , (AI) technology can provide workplace digitalization in order to raise productivity efficiency. (AI) technology will be one kind of production which is replaced by workplace digitalization and it will grow any organization productivity efficiently. Then, (AI) technology will assist overall social economy growth , due to productivity is raised and products can be produced in short time to prepare to sell in consumption market. So, time will be shortened to increase GDP growth fast for the development of (AI) technology countries.

● How can (AI) technology impact to global economic and social and psychological changes?

What will be the development of (AI) technology and predictions concerning the future evolution? The computers and robots will develop conscious, intelligent and minds into humans, enhancing psychological and behavioral abilities and allowing for direct communication with (AI) minds. (AI) technology will be impacted human life by (AI) technology information communicative and environmental influence. A " world brain" and " world mind", this psychological system will be enhanced and enriched the capacities of both individual and collective cognition by (AI) technology of service industries.

(AI) technology with influence these human needs of service industries changes, such as , biological science, finance, entertainment, business, biological science, transportation, communication military etc. The personal computer evolution, the internet and the world wide web which exploded on the scene, linking business, homes, schools, social organizations which were a completely unpredicted phenomenon to influence human life. Kurzweil (1999) predicts that by 2029 year, most human communication will be with machines. According to Person, by 2100 year, there will be human machine convergence.

How can (AI) technology influence environmental protection to make benefits to farming economic growth? (AI) technology can be applied to predict how to solve environmental pollution challenge to avoid to damage any crop or vegetable or rice or fruit etc. food growth. Because environmental experts can gather global environmental pollution data from an environmental database to build a perform a systematic analysis from (AI) technology. The first step is this broad analysis can include understanding, statistical and data gathering techniques to obtain the relevant data, the correlation among the variables involved, and a list of possible models. The next step is to select a set of methods and models that cover all kinds of knowledge and functionalities needed for the decision making process. Once the models are selected, they must be fully implemented by means of machine learning , data mining, statistical or numerical technique. After that, those models must be integrated to build the whole EDSS. The EDSS must be tested to check its performance, accuracy, usefulness and reliability, both from the user's and (AI) technology/computer scientist's point of view. If these is any wrong feature in any development stage, such as model's integration, models' implementation, selection of models, database, problem analysis etc. the developers must come back in the update th required components. When the evaluation phase is all right, the EDSS is ready to be applied to the environment. The great contribution of artificial intelligence to EDSS the integration of several methods complementing the classical statistical models/simulation , statistical analysis, linear models, etc. and numerical models ( control algorithms, optimization techniques etc.) .

This cooperation makes the resulting systems more reliable and powerful in coping with real world environment systems. Date interpretation has been a principal area of research in (AI) technology since the very beginning. The most demanding problem in the environmental assessment context. Knowledge representation permits the definition of the different types of data that the existing methods adapt to the process. There is also a lot of work to clean, repair and transform the huge available quantities of raw data. Apart from this, the availability of meta-information or background knowledge is required to guide the process. Data mining is multi-disciplinary: It covers expert systems, data based technology, statistics, data visualization and unsupervised machine learning. These techniques operate at the level of data and background information, where numerous and often incompatible new commensurate pieces of information from disparate sources have to be brought together ( K, Fedra, 1994). So, it seems that in the future, (AI) technology with the increasing maturity in particular those related to knowledge and engineering, new dimensions can be assisted to users in environmental decision making are available. For example, many environmental systems are characterized both by incomplete models and by limited data. Hence, in the future, (AI) technology will be applied to predict climate change to reduce crop or fruit etc. food agriculture challenge by climate change bad influence.

- Will (AI) technology influence digital economy change to manufacturing industry ?

To understand how the manufacturing business must adapt to prosper in the technology, we need to understand how (AI) technology will change us to shape our daily habits to satisfy our expectation of products to how we shop and even the immediate of the entire process. For example, taxi services are in the crosshairs as on demand transportation services like, available of the touch of a smart phone button expand. In fact, Yellow lab, US country , san Francisco city's largest taxi company is filing for bankruptcy as the industry starts to change faster than almost anyone expected. However, at this point, its more than an app that is changing, some our taxi passengers renting taxi transportation to catch consumption behavior.

(AI) technology will influence digital economy for taxi passenger's individual customer experience, offering a growing renting taxi to catch of service and feedback opportunities when any one taxi passenger who chooses to use mobile phone app online tool to prepaid to rent any taxi more easily.

Also in the long term, (AI) technology can influence vehicles drive themselves of behavior. Already, companies like Google and GM are working on projects to bring fleets of autonomous vehicles to cities at the path of a button.

Moreover, this on-demand service model is beginning to appear across a much broader range of markets. For example , Amazon company is investing in its own fleet of trucks, planes and even drone at the same time as it pushes for same-day delivery of products. As some point, vehicles will be autonomous too. So, it seems that (AI) technique will influence any transportations choose to use digital autonomous driving technology in the future . For Amazon company case, it is not stopping of logistics. It is also aiming to automatically manage the supply of consumer home products with its recently launched Amazon replenishment service, Dash. Dash is a digital service that enables that connected derive to automatically order physical products from Amazon when supplies are running low. So, it seems (AI) technology will be applied to logistic function by digital technology method introduction in the future.

Hence autonomous vehicles will optimize industry supply chains and logistics operations through increased efficiency and flexibility. In fact,

fully automated and lean supply chains will keep reduce load sizes and inventory by leveraging smart distribution technologies and smaller autonomous vehicles by machine man assistance. If Amazon continues to grow market share for online sales by reducing effort required by the consumer to place an order, when also contributing the almost immediate delivery of products to the doorstep. So, it will further fuel the trend toward on-demand derive. As Amazon company fuels the on-demand economy, consumers will expect immediacy in more parts of the digital economy. On top of speed, consumers increasing expect more personalization options.
So, (AI) technology will influence digital manufacturing, such as Amazon publishing to monitor every aspect of every process in real -time and communicating to self-optimized deep learning robotics, new methods of high volume and high customization will become possible. Then, as products merge into product platforms and even services, manufacturers have the opportunity to provide components and platforms used by smaller players. So, (AI) technology will influence manufacturing industry to choose automated SMI lines, robots installed, automation engineers.

Another future (AI) technology development can be applied to space science aspect, such as Automation engineering space in manufacturing process to achieve digital manufacturing benefits to any businesses in the future. Such as reducing cost, shortening manufacturing time, raising efficiency, shortening delivery products to client individual time. How can artificial intelligence give the need and advanced fast and evaluation methods benefits for space exploration? When US NASA ( space exploration organization) achieves any space exploration missions, it will answer this question:
When is it useful to have a machine use (AI) technology to achieve a decision? After all, after millions of years of space exploration and rough 10,000 years of civilization, humans are usually quite good at making decisions in complex uncertain environments. Through, Johns Hoplains University's Applied Physical Lab. Research in (AI) technology enabled systems, which has identified three general use cases for (AI) technology to explore space mission:
First, for some tasks (AI) technology is more cost effectiveness than human. Second, (AI) technology is better suited than humans at solving some, but not all problems. Third, (AI) technology allows NASA organization's space exploration mission to develop machines that ate capable of responding faster than when a human is in the decision loop ( D. Scheidt, 2012, A.

Castano et. al. 2008).

So, the use of (AI) technology to enable science by observing the pace of rapidly evolving phenomena was demonstrated. It is more effectively coordinating and (AI) technology utilizing to earn economic benefits to use for space exploration mission.

However, (AI) technology also have current risk for space exploration. Today (AI) technology is immature and requires further development to reach its potential. For instance, the (AI) technology algorithms that detected the dust derive could not have identified whether the Martain weather represented a threat to the cover. Also it can not yet use instrument input to determine what, where and how to autonomously make the next space science measurement. An equally important factor limiting (AI)'s deployment is that lacks the methodology and technology to effectively test (AI) technology. So, the challenge will testing (AI) enabled system is how (AI) performance can be measured. It would be NASA organization's difficulty to find (AI) technology to develop to carry on researching any space exploration missions in the future. However, (AI) technology will be a good economic benefit choice for space exploration mission in the future.

2.1 What is artificial intelligence potential
benefits and ethical considerations?

The ability of (AI) technology systems to transform vast amounts of complex information into insight has the potential to help solve manufacturing or service challenges for human needs. However, to reap the societal benefits of (AI) systems, humans will need to trust then and make sure that which follow the same ethical principles, moral values, professional codes and social norms that we humans would follow in the same scenario, research and educational efforts as well as carefully designed regulation in order to achieve the most effort of economic benefits goals. For example, international business machines corporation (IBM) is actively engaged both competitors , in global discussions about how to make (AI) ethical and as beneficial as possible for people as social economic benefits.

(AI) is usually defined as the " capability of a computer program to perform tasks or reasoning processes " that human usually associate to intelligence in a human being. Often, it has to do with the ability to make a good decision, even when there is uncertainty, too much information to handle. As an example, play chess or complex card games of entertainment activities is believed to need some form of intelligence in a human being, as well as choosing the best medical facilities in a difficult medical case, or creating something new, such as mathematical theorem or even some form of act, or even driving automatic machine man ( self driving vehicle) replacing human driving in the middle of a crowded city.

(AI) needs depends on what we consider being intelligence in the behavior of a human being act a certain point in time. If human belief about human intelligence changes and we don't believe any longer that a certain task requires intelligence, then a computer program performing that task is no longer part of (AI), it becomes just another boring computer program. So, it means that (AI) technology will replace some old computer programs, if human can invent new generation of (AI) software for any functions or activities to satisfy human needs.

As IBM, it argues intelligence. This means that we aim to build systems that enhance and scale human expertise and skills rather than replacing them. We therefore focus on practical applications of (AI) capabilities that assist people in performing well-defined tasks of needs by exploiting and wide range of (AI)-based services. We also use the term " cognitive computing" it is mean a comprehensive net of capabilities based on technology. It comprises the fields of machine learning, reasoning and decision technologies, language, speech and vision recognition and processing technologies, high performance and high efficient functions for any industries or individual consumers needs. For example, robotics, which are usually very good at doing what which are supposed to in any environment, much have public shopping center, factory etc. places which need simply services from the robot ( machine man), such as cleans the floor of our houses to the robot that can work together with humans in production chains, passing through the warehouse, robots can take care of the tasks of an entire warehouse and the companion robots like Nao, Pepper, Aibo and Giraff, who can entertain use, talk to use and help elderly people to stay connected to their friends, relatives and doctors.

Google company is building automatic machine ( self-driving cars ) and has acquired more than 10 robotics companies. Facebook had opened whole new research facility only on (AI) research. Apply computer has developed Siri. Microsoft computer company has built a similar personalized assistant. Google has Deep mind, a UK company whose long term aim is to build general (AI) and has already great potential to win game to the world champion and IBM is investing a huge amount of resources in applying its Watson cognitive computing system to the medical domains to finance and to personalized education. In Europe, IBM is establishing new centers

in Munich and Milan focused in the application of cognitive computer capabilities to the internet of things and healthcare respectively.
For example, automatic machine man ( self-driving cars) are all about (AI), which used to be able to see what happens in the street ( signals ,lanes, other cars, pedestrians, traffic lights, which need to able predict what other cars and pedestrians will do, and who need to be able to cope with unforeseen situations. Since, most car accidents are due to human fault, it is estimated that the adoption of self-driving cars will save about half of the lives that are usually last in car accidents.
IBM Watson company has to understand spoken language, make sense of massive amount to text , respond correctly to questions in many categories, as well as assess its own confidence in responding to such questions. In the future, (AI) technology can own question/answering capabilities that would be very useful, for example, in assisting a doctor when trying to some to the correct diagnosis for a patient and to propose the best therapy .
Intelligent machines can also rely on huge amounts of data to be used to learn how to make better decisions. This data comes from all of us over the years Facebook users have uploaded more than 250 billion pictures and every day who upload about 350 million more. Every second, we submit 40,000 google search queries. So, (AI) technology will be connected through the web from appliances to traffic lights from cars to watches. Other tasks that are very easy for humans are physical and manipulation tasks, such as walking , running, picking up an object to make its shape and location, restricted environment. But (AI) machine man technology still not able to have the general physical and manipulation capabilities even of a 6 year old.
So, it brings this question: Why do (AI) scientists need to concern ethics? Because (AI) technology is complex, information into insight has the potential to reveal long held secrets and help solve some of the world's most difficult problems. (AI) systems can potentially be used to help discover insights to treat disease, predict the whether, and manage the global economy. So, ethic issues is important to and (AI) scientists . If any one new (AI) technology research investigation could success, it will be a secret to and the (AI) scientists can not permit to their loyalty to any competitors to damage the fair (AI) technology products trading market. The country ( countries) (AI) technology scientists need to concern ethic issues, who need to keep secrets for their countries economic or/and social benefits. This is moral issues to any countries/country loyalty is whose countries

intangible assets. They can not sell (AI) loyalty to any their countries to assist whose economic benefits immorally.

How can (AI) technology influence to global
health care economy development?

According to (AI) lecturer analysis, when combined key clinical health (AI) application can potentially create $150 billion in annual savings for the US healthcare economy by 2026 year. (AI) technology is re-winning modern conception of healthcare delivery. It enables machines to sense, comprehend, act and learn. So which can perform administrative and clinical healthcare functions (Accenture, 2017).

It will help health care service organizations to reduce health care cost, will improve and raise service quality and access. So, (AI) health market size will be predicted growth. (AI) applications in health care include robot-assisted surgery, virtual nursing assistant, administrative workflow assistant, fraud detection, error reduction connected machines, clinical trial participant identifier, preliminary diagnosis, automated image diagnosis and cybersecurity.

What kind of benefits (AI) technology can contribute to healthcare service? (AI) technology can deliver what many health care organizations need, such as financial and operational of labor costs, digital expectations from patient consumers how to use (AI) technology to solve interoperability challenges in any healthcare organizations. Also (AI) technology can be applied to wellness an d lifestyle management, diagnostics, delivers financially but also way of organizational and workflow improvement. So, (AI) technology will be continue to become most prevalent and adoption to healthcare organizations , which must need to enhance structure to be position to take full advantages of new (AI) technological capabilities. (AI) technology can change the nature of work and employment is rapidly changing to make the best use of both humans and (AI) talent in healthcare industry in the future. For example, (AI) technology offers a way to fill in gaps and the rising labor shortage in healthcare. According to Accenture analysis, the physicians shortage is increasing. However, (AI) technology will manufacture healthcare machine men to replace physicians in future one day(2017). Hence, (AI) technology will be invented to raise health care service staffs work efficiency and performance in any hospitals or clinics in the future.

In conclusion, (AI) technology will raise efficiency for any service or manufacturing industries in the future, although, it is possible that it will also rise low skillful workers unemployment numbers. But, the most important influence to human technological innovation will be risen and it will influence human life will be changed to be better, e.g. self drive cars, health care physician machine men, machine man cleaners etc. intelligent machine men will be manufactured to serve for our daily life. Furthermore, (AI) technological products will influence countries trading, some low technological development countries manufacturing businessmen can choose to buy any (AI) products to raise whose productivity and efficiency and reducing cost to achieve economic cost saving result. Also, GDP of trading growth income will increase to the (AI) products sale countries. Hence, it will be beneficial to economic development to both developed and developing countries both in the future as well as (AI) scientists time and money spending will be valued to continue to invest (AI) technology development for human life and economy benefits for long term.

In conclusion (AI) technology will raise macro economy growth and it can create many (AI) jobs , but it also raise the low level technological worker unemployment change. In the future, (AI) technology can be applied to digital technology to attempt to invent any new undiscovered (AI) and digital technology. So, it needs any scientists to continue to research how digital and (AI) technology can be mixed to satisfy human's future undiscovered needs.

2.2Artificial intelligence education development
or the future of defense choice

Nowadays, artificial intelligence (AI) is widely knowledge to be one kind of the dramatic technology. However, it is expected to continue, to have a disruptive impact on human's private and public life, so defense and security will be no exception. But how exactly will these be affected ? How will (AI) defense and security is incremental in nature?

To research why artificial intelligence (AI) has possible to be used to cause autonomous weapons by human. We need to understand these three aspects of relationship. They include cybersecurity and artificial intelligence and machine learning and autonomous weapon systems relationship between of them.

Firstly, we need to know what is the mean of artificial intelligence and cyber defense/offense? It means defense of critical networks: real time, pattern finding, anomaly seeking, it must utilize machine (AI) learning algorithms to efficiently, and instantaneously respond to potential network threats as well as it means human on or out of the loop. On the loop : it means anomaly detection: human notified, IT analysis, response. Out of the loop: it means anomaly detection: (AI) decides best method of response: quarantine, honey pot monitoring, hack-back. Thus, it is possible that (AI) can be used , such as autonomous cyber weapon.

What is artificial intelligence and autonomous weapons? Autonomous weapons mean one kind of weapon that can be selected and engaged a target, without intervention by a human operator. Are these machines artificially intelligent? I believe the answer is not, because present weapons systems are not capable of human level reasoning. But, (AI) algorithms are presently employed to process sensor data, monitor system health, take and respond to vocal commands manage data, navigate. This, future autonomous weapons systems will require stronger (AI) to be secure and operationally and cost effective. Moreover, self-aware autonomous cyber systems are crucial.

What is cybersecurity mean? It means the ability to control access to networked systems and the information they contain. It is acted to prevent , detect, recover, react. It is application objects concern people, process, technology and it's application goals are confidentiality, integrity and popular availability. Thus, what is cyber weapon mean? Walware means viruses, Trojans, zero-days, worms ransomware, spyware etc. Does it require a particular objective? E.g. military paramilitary or intelligence. Does it require physical harm? E.g. functional harm or interruption? Mental harm? Is (AI) a technological weapon that it is an object or tool? What about when it is an weapon agent?

In simplicity, (AI) can be one of scientific weapons platform. When one day, it is invented to be applied to control war planes to fly to any countries to attack enemies or it is invented to be seemed to human to replace soldiers to bring guns or any weapons go to other countries to attack. So, it is possible that future any war defense planes, (AI) technological automatic control weapon can be replaced of human soldiers or war plane pilots to control any war defense planes to go to different enemy countries to attack them easily. It is very horror matter to threaten global human's ourselves life in the future , if (AI) automatic control war defense planes or (AI) automatic control machine soldiers were invented successfully.

Hence , when (AI) can be applied to weapons platforms, it structures that launch weapons, i.e. jets, ships, vehicles. (AI) platform and weapon and software architecture components are be done one (AI) technological weapons systems. Thus, human will encounter any (AI) benefits or risks

( threats) causes in the same time as soon as possible. If we can predict when (AI) weapon system will be manufactured or invented successfully. Then, we can reduce (AI) weapon systems risks , if we can threaten any (AI) scientists continue to invent any undiscovered (AI) weapons in any time to avoid the future first time (AI) weapon war occurrence in possible.

The (AI) weapon system risk means autonomy: the ability to problem solve technological war , when (AI) weapon system is manufactured successfully, the power to act, how to damage the (AI) weapon system. The power to chance to stop (AI) weapon system manufacturing processes, ability to create a new goals, how to change the (AI) weapon system inventors' or scientists' minds to avoid to apply (AI) tools to achieve attack goals to change to another positive goal. Due to human can't know a prior what an autonomous (AI) weapon system will do.

Although, human is known what (AI) is , but human is also known when (AI) scientists whose emergent behaviors will do to change to do any negative behaviors from positive behaviors. Whatever (AI) weapon system design we use, there will be cybersecurity, problems arising from computation design/complexity. Due to any one (AI) scientist can manipulate the system to act against itself, or who can utilize traditional " cyber weapons" against the (AI) weapon system, or who can manipulate the system to lie to humans, but also due to complexity, there is no way to know if it is lying or not or bounded rationality : satisficing.

Finally, the most serious (AI) technological invention risks are human is unknown these aspects of (AI) absolutely: They are not simple automatic systems, learning reasoning, communication of " self-aware" systems. Thus, human will face (AI) technological invention risks or threats. We need to find any methods to avoid (AI) weapon system is manufactured successfully to avoid (AI) technological war can occur in future anyone day. Consequently, why (AI) scientists do not choose to invent (AI) machine men own human reading, writing, speaking, judgement, analyizing abilities to develop human's future education industry, but they choose to invent (AI) machine men own human weapon to attack enemy ability.

Hence, (AI) scientists have moral responsibilities to avoid to invent (AI) machine men to own human's attack abilities. I shall indicate future what aspects of (AI) machine men can be applied to different education industry aspects as below:

2.3 Online technology and online book technology influences artificial intelligence mind development

Nowadays, online technological invention bring online book technological development. Also, artificial intelligent technological machine men had been invented to link internet to do any jobs, e.g. children can find any data from artificial intelligent machine men when the artificial intelligent machine man had been installed internet and computer function, then children can find any online books to read from the artificial intelligent machine man. Such as Japan artificial intellgent machine men had installed computer and internet function, the Japan family children can find any online books to read from the artificial intelligent machine man at Japan any families‘ homes conveniently. Hence, it implies that future one day, artificial intelligent machine has possible to be invented to own human's reading and/or writing abilities.

For example,online book publishing is one kind of popular internet technology. For example, Amazon publish is as a business model with many potential advantages, relative to a physical operation. It held out the potential of lower book inventing and distribution costs and reduced overhead. Consumers could find the books, they were looking for more easily and a variety book topic choices could be offered for sale. It can accept and fulfill orders from almost any domestic location with equal ease. And most purchasers made on its site would be exempt from sales tax. One Amazon strategy hand, it would have to make its returns and redress processes transparent and reliable, and offer other ways for clients to learn, as much about the book possible before buying. Future online book market development trend, such as Amazon, Barnes & Noble etc. online book shops.

Hence, online book store technology can be applied to artificial intelligent technology. Such as artificial intelligent machine men can apply computer technology to learn the abilities of reading and/or writing any books either on paper or on computer. Hence, it is possible that artificial intelligent machine men will have similar human's writing and/or reading books ability when they own human's mind ability. However, it bring this questions: Can artificial intelligent machine men own human's mind abilities? If they own human's mind abilities, is it mean that they can write and/or read any books? Can artificial intelligent machine men own human's mind abilities to create to write any books? Can artificial intelligent machine men own human's mind abilities to read and make any judgements or decisions more accurate than human's judgements or decisions? To answer these questions? I shall indicate that online book reading and writing technology can be applied to artificial intelligent machine men reading and writing technology. Because they are similiar computer mind technological development. So, I believe that future artificial intelligence machine men can be invented to own similar human's reading and writing's mind abilities in future one day.

I believe artificial intelligence and online technological reading abilities are very similiar. Nowadays, computer can be invented to attempt to read and write any books by human. Why can not artificial intelligent machine men replace computer to read and write any books? Artificial intelligent machine men can replace human to attempt to write or/and read books, due to artificial intelligent machine men had invented to own human mind to do some jobs and their mind had been invented to be similiar to human behavioral abilities to do these behaviors, e.g. cooking, driving, playing games, singing songs, speaking, listening, frighting etc. different human's abilities. So, it seems that artificial intelligent will be possible to be invented to own human's mind abilities to do any writing or reading behaviors or functions.

## 2.4 Prediction of artificial intelligence reading and writing abilities

development

What is future trend of artificial intelligence reading and writing abilities development? To answer this question, we need to know what benefits of artificial intelligent machine men can attribute to human's needs when they can own any human's mind to read or/and write any books.

I shall indicate e-books reading and writing example, if artificial intelligent machine men can be invented to own human's mind to write and/or read e-books on computer. Then, it brings this question: Can artificial intelligent machine men assist human to learn to do judgement to solve any challenges?

I believe that when artificial intelligent machine men can be invented to own human mind to write or/and read any books, then they will own human's mind ability to make judgement to solve any challenges more accurately, even their decisions can be more accurate to compare to human's decisions. So, artificial intelligent machine mens' writing and reading ability is the main factor to cause their mind to do any judgement in order to make any decisions more accurately. Consequently, in future one day, artificial intelligent machine mens' writing and reading ability will be invented to similar human's reading and writing abilities as well as their minds can also be invented to similar human's minds as well as their judgement abilities can be invented to similar to human's judgement abilities to make any decisions more accurate.

2.5The influences when AI is invented to own
human's mind and judgement abilities

Finally, I shall discuss what are the influences when AI is invented to own human's mind and judgement abilities in our future job market. The achievement of artificial intelligent (AI) machine men achievement requirement of owning human's mind and judgement abilities which requires extensive manual labor, and by augmenting the calling process with machine learning, the process where speed and accuracy are needed to close to human's mind and judgement abilities. Expert human race callers now have better information at artificial intelligent machine men at their fingertips faster.

Hence, if the above those requirements are achieved to satisfy artificial intelligent machine men ind and judgement abilities demand to close or exceed humans' mind and judgement abilities. Then, I believe that future human's some simple jobs must be replaced by (AI) machine men. Even, human's some professonal jobs, e.g. lawyer, accountant, administator, typing etc. professional skillful jobs, which will be either replaced or will be assisted by (AI) machine men. For example, (AI) machine men learn how to type english or other language words to do typing job ; they can learn how to apply accounting knowledge to record any firm's income and expenditure record of accounting job; they can also learn how to assist architects to design any architectural building drawing plans to do architect jobs; they can learn how to analyze any court evidences to judge any criminal or civil cases and assist lawyers to give legal advices to achieve more reasonable judgement for any legal cases; they can also learn how to assist firm's managers or administrators to manage any organization teams efficiently.

Consequently, when (AI) machine men can be invented to achieve to exceed human's mind and judgement abilities level. Then, I believe that they can do instead of human' simple jobs, which can do even human's more difficult and more judgement requirement of professional skillful jobs. So, (AI) machine men must need to achieve to do any jobs, they are same, even exceed to human professionals' abilities. Then, it will cause a lot of human's jobs to be disappeared or some human's jobs will be replaced by owning judgement and mind abilities of (AI) machine men to do.

Hence, future many human's jobs will be replaced by technological labors. Employers choose to buy (AI) machine men to replace human labors. The reasons include (AI) machine men have none unhappy, angry emotin to influence their low efficiencies and low productivities. Their judgement and mind abilities can exceed human's abilities or do any jobs to compare better performance to human's abilities. Consequently, different occupation labors need to prepare to learn how to co-operate with (AI) machine men to let future employers feel (AI) machine men will be human's assistant to assist human to do jobs efficiently when human and (AI) machine men work together. It aims to avoid future employers feel (AI) machine men's judgement and mind abilities can exceed any low knowledgeable and skilful occupation labors, even high knowledge and skilful occupation labors. It means that (AI) machine men are only labors' assistant if (AI) machine mens' judgement and mind abilities are below under to human labors' judgement and mind abilities.

Consequently, to avoid (AI) machine men can replace human to do any simple or complex jobs to cause any future any occupation labors' competitiors. I recommend that it is right time labors ought prepare to learn different skills. So, every individual labor does not only concentrate on one kind of skill. Because supposing one kind of the occupation labor's job duties are replaced by (AI) machine men. If the employee had owned more than one kind of occupation skill. Then, I believe that who can avoid the unemployment threat more easier than the employee only owned one kind of occupation skill, when (AI) machine men had invented to own human's mind and judgement abilities in future one day. The most important, I believe that (AI) invention will be applied to be teach how to learn human's skills and mind ability. Such as education industy, teacher won't be replaced by (AI), otherwise, (AI) will be teacher's assistant to help them to do education data gather or teaching jobs. So, teachers won't be replaced by (AI), otherwise, teachers will depend on (AI) data gather or teaching job to give them opinions how to solve student's teaching challenges as well as teachers can concentrate on researching education jobs for schools' benefits if (AI) technology can be invented to on human's mind and judgement and reading and writing abilities in the future.

## 2.6Artificial intelligence and the future of defense or teaching choice

Nowadays, artificial intelligence (AI) is widely knowledge to be one kind of the dramatic technology. However, it is expected to continue, to have a disruptive impact on human's private and public life, so defense and security will be no exception. But how exactly will these be affected ? How will (AI) defense and security is incremental in nature? If (AI) technological machine men are applied to teach students in education aspect, is it better to my next generation learning develpment more than they are applied to war attack aspect.

To research why artificial intelligence (AI) has possible to be used to cause autonomous weapons by human. We need to understand these three aspects of relationship. They include cybersecurity and artificial intelligence and machine learning and autonomous weapon systems relationship between of them. Basic on (AI) machine can be invented to learn any new knowledge, so if (AI) machine men are taught how to attack enemy, which will be such as human soldier function. But, if (AI) machine men are taught how to learn university knowledge to teach students. Then, they will be such as human lecturer function. So, when (AI) is invented to own human mind and judgement and learning abilities, then they will be either human's enemy or human's assistant, such as university lecturer's assistant.

Firstly, we need to know what is the mean of artificial intelligence and cyber defense/offense? It means defense of critical networks: real time, pattern finding, anomaly seeking, it must utilize machine (AI) learning algorithms to efficiently, and instantaneously respond to potential network threats as well as it means human on or out of the loop. On the loop : it means anomaly detection: human notified, IT analysis, response. Out of the loop: it means anomaly detection: (AI) decides best method of response: quarantine, honey pot monitoring, hack-back. Thus, it is possible that (AI) can be used , such as autonomous cyber weapon. If (AI) is applied to make the decision best method of response to learning aspect, such as univeristy different subject knowledge. Then, it will be one good technological educational tool to teach university students.

In simplicity, (AI) can be one of scientific weapons platform or one of university teaching tool. When one day, it is invented to be applied to control war planes to fly to any countries to attack enemies or it is invented to be seemed to human to replace soldiers to bring guns or any weapons go to other countries to attack. So, it is possible that future any war defense planes, (AI) technological automatic control weapon can be replaced of human soldiers or war plane pilots to control any war defense planes to go to different enemy countries to attack them easily. It is very horror matter to threaten global human's ourselves life in the future , if (AI) automatic control war defense planes or (AI) automatic control machine soldiers were invented successfully. Otherwise, when one day, (AI) machine lecturer is invented to be applied to learn university different subjects knowledge to replace lecturers to copy lecturer's every prepared lecturer course to speak to let students to listen when they are sitting in university halls as well as the (AI) machine lecturer can make analysis and judgement response to answer every student's enquire immedicately after it had speaking all courses to students to listen in lecturer hall every time. Then, it can let human lecturer does any education job duty, e.g. research education work. So, (AI) machine lecturer will be future human lecturer's assistant in future one day.

Finally, the most serious (AI) technological invention risks are human is unknown these aspects of (AI) absolutely: They are not simple automatic systems, learning reasoning, communication of " self-aware" systems. Thus, human will face (AI) technological invention risks or threats if human invent (AI) machine man to learn how to attack enemy. Otherwise, if human invent (AI) machine man to learn how to teach univerity student. I believe that my future university students can raise learn ability and writing ability and reading ability from (AI) machine lecturer teaching more than human lecturer teaching.

## 2.7Online technology and online book technology influences artificial intelligence mind development

Nowadays, online technological invention bring online book technological development. Also, artificial intelligent technological machine men had been invented to link internet to do any jobs, e.g. children can find any data from artificial intelligent machine men when the artificial intelligent machine man had been installed internet and computer function, then children can find any online books to read from the artificial intelligent machine man. Such as Japan artificial intellgent machine men had installed computer and internet function, the Japan family children can find any online books to read from the artificial intelligent machine man at Japan any families' homes conveniently. Hence, it implies that future one day, artificial intelligent machine has possible to be invented to own human's reading and/or writing abilities.

For example,online book publishing is one kind of popular internet technology. For example, Amazon publish is as a business model with many potential advantages, relative to a physical operation. It held out the potential of lower book inventing and distribution costs and reduced overhead. Consumers could find the books, they were looking for more easily and a variety book topic choices could be offered for sale. It can accept and fulfill orders from almost any domestic location with equal ease. And most purchasers made on its site would be exempt from sales tax. One Amazon strategy hand, it would have to make its returns and redress processes transparent and reliable, and offer other ways for clients to learn, as much about the book possible before buying. Future online book market development trend, such as Amazon, Barnes & Noble etc. online book shops.

Hence, online book store technology can be applied to artificial intelligent technology. Such as artificial intelligent machine men can apply computer technology to learn the abilities of reading and/or writing any books either on paper or on computer. Hence, it is possible that artificial intelligent machine men will have similar human's writing and/or reading books ability when they own human's mind ability. However, it bring this questions: Can artificial intelligent machine men own human's mind abilities? If they own human's mind abilities, is it mean that they can write and/or read any books? Can artificial intelligent machine men own human's mind abilities to create to write any books? Can artificial intelligent machine men own human's mind abilities to read and make any judgements or decisions more accurate than human's judgements or decisions? To answer these questions? I shall indicate that online book reading and writing technology can be applied to artificial intelligent machine men reading and writing technology. Because they are similiar computer mind technological development. So, I believe that future artificial intelligence machine men can be invented to own similar human's reading and writing's mind abilities in future one day.

I believe artificial intelligence and online technological reading abilities are very similiar. Nowadays, computer can be invented to attempt to read and write any books by human. Why can not artificial intelligent machine men replace computer to read and write any books? Artificial intelligent machine men can replace human to attempt to write or/and read books, due to artificial intelligent machine men had invented to own human mind to do some jobs and their mind had been invented to be similiar to human behavioral abilities to do these behaviors, e.g. cooking, driving, playing games, singing songs, speaking, listening, frighting etc. different human's abilities. So, it seems that artificial intelligent will be possible to be invented to own human's mind abilities to do any writing or reading behaviors or functions.

## 2.8Prediction of artificial intelligence reading and writing abilities

development

What is future trend of artificial intelligence reading and writing abilities development? To answer this question, we need to know what benefits of artificial intelligent machine men can attribute to human's needs when they can own any human's mind to read or/and write any books.

I shall indicate e-books reading and writing example, if artificial intelligent machine men can be invented to own human's mind to write and/or read e-books on computer. Then, it brings this question: Can artificial intelligent machine men assist human to learn to do judgement to solve any challenges?

I believe that when artificial intelligent machine men can be invented to own human mind to write or/and read any books, then they will own human's mind ability to make judgement to solve any challenges more accurately, even their decisions can be more accurate to compare to human's decisions. So, artificial intelligent machine mens‘ writing and reading ability is the main factor to cause their mind to do any judgement in order to make any decisions more accurately. Consequently, in future one day, artificial intelligent machine mens' writing and reading ability will be invented to similar human's reading and writing abilities as well as their minds can also be invented to similar human's minds as well as their judgement abilities can be invented to similar to human's judgement abilities to make any decisions more accurate.

## 2.9The influences when AI is invented to own

human's mind and judgement abilities

Finally, I shall discuss what are the influences when AI is invented to own human's mind and judgement abilities in our future job market. The achievement of artificial intelligent (AI) machine men achievement requirement of owning human's mind and judgement abilities which requires extensive manual labor, and by augmenting the calling process with machine learning, the process where speed and accuracy are needed to close to human's mind and judgement abilities. Expert human race callers now have better information at artificial intelligent machine men at their fingertips faster.

Hence, if the above those requirements are achieved to satisfy artificial intelligent machine men ind and judgement abilities demand to close or exceed humans' mind and judgement abilities. Then, I believe that future human's some simple jobs must be replaced by (AI) machine men. Even, human's some professonal jobs, e.g. lawyer, accountant, administator, typing etc. professional skillful jobs, which will be either replaced or will be assisted by (AI) machine men. For example, (AI) machine men learn how to type english or other language words to do typing job ; they can learn how to apply accounting knowledge to record any firm's income and expenditure record of accounting job; they can also learn how to assist architects to design any architectural building drawing plans to do architect jobs; they can learn how to analyze any court evidences to judge any criminal or civil cases and assist lawyers to give legal advices to achieve more reasonable judgement for any legal cases; they can also learn how to assist firm's managers or administrators to manage any organization teams efficiently.

Consequently, when (AI) machine men can be invented to achieve to exceed human's mind and judgement abilities level. Then, I believe that they can do instead of human' simple jobs, which can do even human's more difficult and more judgement requirement of professional skillful jobs. So, (AI) machine men must need to achieve to do any jobs, they are same, even exceed to human professionals' abilities. Then, it will cause a lot of human's jobs to be disappeared or some human's jobs will be replaced by owning judgement and mind abilities of (AI) machine men to do.

Hence, future many human's jobs will be replaced by technological labors. Employers choose to buy (AI) machine men to replace human labors. The reasons include (AI) machine men have none unhappy, angry emotin to influence their low efficiencies and low productivities. Their judgement and mind abilities can exceed human's abilities or do any jobs to compare better performance to human's abilities. Consequently, different occupation labors need to prepare to learn how to co-operate with (AI) machine men to let future employers feel (AI) machine men will be human's assistant to assist human to do jobs efficiently when human and (AI) machine men work together. It aims to avoid future employers feel (AI) machine men's judgement and mind abilities can exceed any low knowledgeable and skilful occupation labors, even high knowledge and skilful occupation labors. It means that (AI) machine men are only labors' assistant if (AI) machine mens‘ judgement and mind abilities are below under to human labors' judgement and mind abilities.

Consequently, to avoid (AI) machine men can replace human to do any simple or complex jobs to cause any future any occupation labors‘ competitiors. I recommend that it is right time labors ought prepare to learn different skills. So, every individual labor does not only concentrate on one kind of skill. Because supposing one kind of the occupation labor's job duties are replaced by (AI) machine men. If the employee had owned more than one kind of occupation skill. Then, I believe that who can avoid the unemployment threat more easier than the employee only owned one kind of occupation skill, when (AI) machine men had invented to own human's mind and judgement abilities in future one day.

### 2.10Why does AI machine lecturer can raise education quality?

When (AI) machine men can own human reading and writing and judgement and analytical abilities, then they can replace university lecturers to teach students to raise students' learning abilities absolutely. Then, it bring this question: Why does I machine lecturer can raise education quality? Why do universities prefer to apply (AI) machine lecturer to teach teachers more than human lectuer in university lecturer hall learning environment? Will (AI) university lecturers replace human lecturers to teach students to learn at university lecturing halls popularly? Can (AI) university lecturers replace university human lecturers to teach students more easily and it can let students feel more easily to learn when they are listening what (AI) university lecturers are teaching to them every time university lecture.

In university today, nearly all students need to attend university lecturing hall to listen their lecturer's teaching in every time course. However, many students do not feel interesting to attend university halls to listen human lecturer's teaching. The reasons include, they are busy, so no time to attend lecturer's hall to listen lecturer's teaching; or they feel bore to listen their lecturer's teaching; they feel difficulty to learn; they have confidence to exam and do their assignments, so they feel that they do not need to go to lecturing halls to listen their human lecturer's teaching. However, if one day, (AI) machine lecturers are invented to teach university students to learn and solve their learning difficulties. Can it raise student individual learning interest, due to (AI) machine lecturers' education quality is better than human lecturers' education quality?

What will influence to university students if (AI) machine lecturer can invented to replace human lecture? The influences will include such as below:

First reason: the only way is going to be useful to university lecturers are if all (AI) machine lecturers are well-informed and fully supported to assist human lectuers to teach whose students to let them to listen whose teaching absolutely. So, human lecturers can concentrate on doing any education research and data gathering jobs to prepare for (AI) machine lecturers to help them to explain human lecturers' every time prepared course contents more efficiently. So, (AI) machine lectuers can help human lecturers to share whose teaching time in lecturing halls. Human lecturers' can spend whose hall lecturing time to do whose educational research or other educational gathering jobs absolutely.

The second reason, the human lecturer (Human capital) has ability and efficiency of concentrating on education data gatehering research jobs to prepare to write whose books. When (AI) machine lecturer replace the human lecturer to spend time to attend lecturing hall to teach students. Fo long term, the human lecturer can raise education productivity growth and education quality raising, due to who only concentrate on searching or gathering data to prepare to write whose books to raise their education level.

In macro and micro economic view, the well (AI) machine lecturer educated labor ( human capital) is often replaced to human lecturer as one of the critical factors to influence rapid education productivities and educational quality growth to the Asia developing countries' any regions or cities. Because any of these Asia developing countries, such as China,

Korea, Philippines etc. countries which need have well educated and knowledgeable lecturer labors to raise any universities' educational productivities and educational qualities growth. So (AI) assistant lecturer factors which ought have close relationship to cause the good or bad future student learning effectiveness and education or learning qualities raising in these any one of Asia developing countries.

The third reason, for the big population of student growth number example, China's student growth rate is larger than school growth rate. If China expect every students have enough chance to study in schools, but university human lectuer numbers are not enough to supply to universities to teach their students. I believe that (AI) machine lecturer is only one kind of teaching method to solve these big population countries' university lectuer number shortage challenge.

In conclusion, in long term, (AI) machine lecturers can solve university human lecturer shortage challenge as well as they can attract many students to attend lecturing halls and human lecturers can raise education quality when they can concentrate on searching or gathering data to prepare their education career, when (AI) machine lectuers replace them to spend time to attend univesity halls to teach students in every university lecturing time.

2.11 Future AI machine education market

I believe that when AI (artificial intelligent machine men) which can invented to own to similar to human mind, learning, language, analytical, judgement abilites. Then, which can be applied to any education market service industy. (AI) potential education market service industy includes such as below:

2.12(AI) university lecture assistant

Future (AI) machine men can assist univerity lectuers to attend university halls to attempt to teach university students for different subjects, e.g. english, math, economic, math, engineering, art, architect etc. different subjects. It depends on the human lectuter who prepares to spend time to teach the (AI) machine lecturer to remember whose teaching subject. For example, the economic lecturer spend one year time to teach the (AI) machine lecturer to learn all economic knowledge. Then, the (AI) machine lecturer can use its machine brain to remember all the human lecturer's economic concepts and prepared teaching economic contents within the one year. Hence, after one year the (AI) machine lecturer can remember all the human lecturer's economic concepts and economic theories and economic contents to prepare to attend university lecturing halls to teach all first year undergraduated first year economic students confidently. It means that the human lecturer's job duties will change to teach (AI) machine lecturer to learn whose economic knowledge to prepare to let the (AI) machine lecturer to replace whom to teach whose university students; so the human lectuer can spend more time to do other research job for whose university education development. Hence, the (AI) machine lecturer can share the human lecturer teaching job as well as the human lectuer can concentrate on spending time to do whose research jobs for whose university education development. This is one both win strategy to university and the lecturer if (AI) machine lecturer is invented to assist future university lecturer's teaching jobs.

2.12.1(AI) secondary and primary teacher assistant

In the future (AI) technolgical development, instead of (AI) machine men can be applied to university education aspect. Future (AI) machine men can also be applied to secondary and primary teaching aspect. For example, primary and secondary schools do not need attend classroom to teach students. (AI) machine teachers can replace them to attempt to do teaching job. They only need to spend one year time to prepare to teach (AI) machine teacher to learn how to apply their teaching skill concern their subjects who need to teach to their students, e.g. english language writing and reading and spelling skill, sing song skill, drawing picture skill, calculation skill etc. different studying skill. Then, the (AI) primary or secondary machine teacher can apply the primary or sendary human teacher skills to attempt yo teach whose students. Hence, the primary and secondary human teacher whose duties will change to learn how to teach whose teaching skills to let the (AI) primary or seondary machine lecturer to remember how to apply human skills to teach whose students for different subjects, such as, english writing and reading and spelling language skills, singing songs language skills, math calculation skills etc. Hence, future primary or secondary school teachers who responsibilities will change to learn how to teach (AI) machine teacher teaching skills to prepare to replace them to teach their students in classroom.

2.12.2 (AI) scientific research assistant

Future (AI) machine men can be applied to science research aspect, instead of school education job. For example, (AI) machine men can be any scientist's assistant, e.g. space scientist, earth or ocean scientist, human or animal behavioral psychological scientist, climate scientist, chemical scientist, drug scientist etc. How can (AI) machine men can be any kind of scientist to assist scientists to do research jobs ? I shall indiate such as below: For space science example, the (AI) machine space scientist can assist human space scientist to gather space data to assist space scientist to research any undiscovered material to cause our earth, even space. Hence, the space scientist only need to teach the (AI) machine scientist to learn how to help them to apply space technological tools to gather data and then enter all data to computer to record, even the (AI) machine scientist can store all space data discovered record to their machine brain every day. Hence, the human space scientist does not need to spend much time to do gathering data job. The (AI) machine space scientist can help whom to do these space data gathering job, then the human space scientist can concentrate on spendin time to do space research job in whose space science laboratory every day.

For earth science example, the (AI) machine earth scientist can help the earth scientist to go to anywhere to gather earth or ocean natural activity data in our earth every day. Then, the earth scientist only need sit in whose earth laboratory to wait the (AI) machine earth scientist to come to whose laboratory to give whose gathering every day earth or ocean natural activity data to do future research job. Hence, the earth scientist does not need to leave whose laboratory to do any data gathing jobs concern earth or ocean natural activities. The (AI) machine earth or ocean scientist had helped him/her to go to our earth or ocean anywhere to do any earth or ocean activities data gathering jobs every day. Hence, the earth or ocean scientist can concentrate on spending whose time to do any research jobs in laboratory. It means that the (AI) machine earch or ocean scientist had replaced whom to do all outdoor original gathering data jobs.

For these human or animal behavioral psychological scientist, climate scientist, chemical or drug scientist, scientist all examples, the (AI) machine human or animal behavioral psychological scientist can help them to do any data gathering job, e.g. the (AI) machine scientist can learn how to help human or animal behavioral psychological scientist to contact human or animal to observe their daily activities and record all their activities data to transfer all these daily activites data to let the human or animal behavioral psychological scientist to do psychological researching analysis only. The (AI) machine climate scientist can help the human climate scientist to arrive anywhere to observe climate changes and record climate changes daily. Then, the human climate scientist only need to wait the (AI) machine climate scientist's gathering climate change data record from whose machine brain to do climate changing predict research job in climate laboratory every day. The (AI) chemical or drug machine scientist can help the drug or chemical scientist to gather data of new drug or chemical from internet channel every day. So, the human chemical or drug scientist only need to do researching job after the (AI) machine scientist transfers all daily chemical or drug information to let them to know from internet channel. It means that the chemical or drug human scientist does not need to spend much time to gather drug or chemical new data development trend from internet. The (AI) machine chemical or drug scientist had helped them to do data gathering job every day.

Consequently, future (AI) machine men can do education and research aspects of jobs duties and their role are only human scientists or primary or secondary teachers or university lecturers whose assistants either to share scientist's data gathering job or share teachers or lecturers' teaching job.

2.12.3Artificia intelligent future education market development

Artificial intelligent robots invention

negative impacts

Indeed, artificial intelligence, computing can learn to something that effectively reasons, thinks, if (AI) learns more powerful and valuable complement to human capabilities; improving medical diaguoses, weather prediction, supply-chain management, transportation and even personal choices about where to go on vacation or what to buy, how to learn teaching any subjects knowledge to assist school teachers to teach students, e.g. accounting, law, architecture, language etc. subjects knowledge It can bring benefits to human it (AI) robots can learn teacher's any subjects to teach students or skillful knowledge, e.g. driving, taking care old people at home, manufacturing vehicles in factories. Otherwise, if (AI) robots learn how to be applied to be weapons to attack enemy. It will bring harm to human's life safety. As a result, technology can assist human's development when it can be applied to do meaning jobs, but it can also damage human's development when it can be applied to do harmful human's behaviour, e.g. attacking enemy to cause robot technological war. Consequently, the negactive impact will be depending on how humans teach (AI) robots to learn when some humans intends to teach (AI) robots to attack enemy. Thus, scientists need to know to educate (AI) robots to learn positive knowledge to attribute to us to raise our standard of living, if it is a tool in the service of humans, making our lives better. Otherwise, who ought not educate (AI) robots to learn negative knowledge to attribute to influence our's life safety when they are applied to attack enemies.

However, some scientists believe (AI) robots will have negative impact to influence our economy and society. Such as (AI) robots will destroy most jobs, it will make humans foolish, due to humans depend on (AI) robots to assist us to do any jobs; it will destroy people's privacy and it will enable bias and abuse and it will eventally exterminate humanity. Hence, when time comes to develop computers really think, intelligence is the same things as consciousness, the brain is a computer. Then, scientists must have responsibilities to concern how to teach (AI) robots to learn useful or attributable human's knowledge, not harmful human's knowledge to avoid future invented (AI) robots are taught to learn how to attack ourselves.

2.12.4Future AI tutoring system potential
development market

Humans can learn (AI) robots how to educate our next generation after it has be taught any knowledge from human. Thus, it brings this question: How can human teach new knowledge to (AI) robots to learn successfully? I shall indicate some scientists‘ evidences to explain that it is possible (AI) robots had abilities to learn human's mind, judgement and analytical abilities, reading, writing, speaking abilities in future one day absolutely. I shall indicate what is intelligent tutoring systems (ITS) example as below:

When are computer-based tutors which act as a supplement to human teachers. The major advantage of an (ITS) is, it can provide personalized instructions to students according to their cognitive abilities. In this scenario, an intelligent tutoring system can be quite relevant to solve unavailability of skilled teachers challenges. Such as India has many students who demend teachers to learn to them, but India is facing skilled teachers shortage challenge. (ITS) are computer-based tutors, which act as a supplement to human teachers. An intelligent tutoring system is educational software containing an artificial intelligence component. The software tracks students work, tailoring feedback and hints along the way. By contenting information on a particular student's performance, the software can make interences about strengths and weaknesses, and can suggest addition work.

The ITS's one of main advantages is individualized instruction delivery to the group of same learning ability of student in evey classroom, which means the system ill adapt itself to different categories of students. A real classroom is usually heterogeneous where these are different kinds of students, from slow learners to fast learners. It is not possible to provide attention to them individually. Thus, the teaching may not b beneficial to all students. An ITS can eliminate this problem, because in this virtual learning environment, the tutor and the student has a one-to-one relationship. When the school applies TIS intelligent tutoring systems to teach its students, the students can learn in whose own method. Another advantage is their using this system teaching can be accomplished with there is trained teachers. Hence, the functionalities of this intelligent tutor system can be divided into three major tasks: (1) organization of the domain knowledge, (2) keeping track of the knowledge and peformance of the learner, (3) planning the teaching strategies on the basis of the learner's knowledge state. Hence, to carry out these tasks, the ITS have different modules and interfaces for communication ( Wenger, 1987; Freedman, 2000; Chou et. at, 2003).

The criteria of robots in education include domain of the learning activity, location of the activity, the role of the robot, types of robots and types of robotic behavior. Scientists indicate that robots are primarily used to provide language, science or technology education and that a robot can take in the role of tutor, tool or peer in the learning activity. However, robotics can include these kinds, such as social robotics, pedagogy, human robot interaction and educational robotics.

Early, industrial robots are invented to apply to manufacturing industry only in 2008 beginning. robots are slowly beginning a process to everyday, lives both at home and at school (IFR, 2008). Nowadays, the most popular countries accept to apply robots to replace human's some jobs, include Japan, Korea, USA, Australia, Germany, Holland.

What is the domain of the learning activity to robots? To develop robots to education industry, the first criterion of the two main categories are robotics and computer educaion ( the awareness of technology that could be referred as technical education) and non-technical education ( science and language). The technical education means giving students the knowledge of robots and technology. For example, introducing computer science and programming and to familiarize undergraduate students with technology and schol students were gradually exposed to technical subjects using robots. A lesson plan usually involves first an initial introduction to programming the robot ( introduction phase) and then the students apply their knowledge practically by making their robots work ( intensive phase. The second observed domain in the area of robots in education are non-robots in education are non-technical subjects (such as the sciences), where schools witness the employment of robots as an intermediate tool to impact some form of education to students in classrooms, such as mathematics. The third common domain is the use of robots to teach a second language. For example, English was taught to Asia countries children by robots in by researchers from the robotics laboratory. For another example, the implication of using robots to teach a second language have been well documented by computer science researchers in Taiwan, where it is stated that children are not as hesitant to speak to robots in a foreign language as they are talking to a human instructor. So, it seems, future robots can be a language teachers to the learning foreign language students. However, the language robots require having accurage speech recognition ability and how to learn in acknowledging the use of robots for language instruction ability in future robots language education market.

2.12.5How can potential social teaching robots
assist to teacher in school?

What benefits do (AI) robots provide to education industry? Artificial intelligence can indicate how to improve courses, when teachers may no always be aware of gaps in their lecturers and educational materials that can let students confused about certain concepts. Different students have different learning styles, abilities, interests and needs. For classroom situation example, on teacher in a classroom of 20 to 30 students will rarely be able to cater to each of those needs. Homework and classes could be customized based on a student profile, interests can be cultivated based enhanced by exposing students to different course and content.

Artificial intelligence can offer a way to solve that problem. For online or non-online course providers cases, will have already benefits if these apply (AI) robots to educate. When a large number of students are found to submit the wrong answer to a homework assignment, the system alerts the teacher and goves future students a customized message that offers hints to the correct answer.

This type of system helps to fill in the gaps in explanation that can occue in online or non-online courses, and helps to ensure that all students are building the same conceptual foundation. Rather than waiting to either hear back from the professor in lecturng hall or classroom from classroom education method or receive back professor's feedback message from the internet online education channel. So, classroom students or online learning students can get immediate feeback that helps them to understand a concept and remember how to do it correctly the next time around if (AI) robots can be applied to assist university professor or primary/

secondary teachers to teach whose students either in classroom teaching learning environment or online teaching learning environment. Thus, (AI) educational and social robots can be attempted to accepted to teach primary school, high school or university students.

How to apply (AI) education robots to teach students in which different educational functional aspects?

Firstly, on grading system hand, artificial intelligence can automate basic activities in education, like grading. In, college, grading homework and tests for large lecture courses can be for much grading activites to secondary teachers or university lecturers to spend much time to do grading jobs for every student. Even in lower grades, teachers often find that grading takes up a significant amount of time, time that could be used to interact with students, prepare for class, or work on professional development. When, (AI) may not every be able to truly replace human grading, it's getting pretty close. it's now possible for teachers to automate grading for nearly all kinds of multiple choice and fill-in-the blank testing and automated grading of student writing may not be far behind. Today, essay grading software will improve over the coming years, allowing teachers to focus more on in class activities and student interface then grading.

Secondly, on (AI) tutor's support hand, students could get additional support from (AI) tutors. When, there are obviously things that human tutors can offer that machines can't. The future could see more students being tutored by tutors that only exist in zero and ones. Some tutoring programs based on artificial intelligence already exist and can help students through basic, mathematics, writing, accounting, law and other subjects. (AI) tutors can teach students fundamentals for helping students learn high-order thinking and creativity, something that real-world teachers are still required to facilitate. It should not rule out the possibility of (AI) tutors being able to these things in the future. With the rapid technology advancement that has marked the past few decades, advanced (AI) tutoring systems will be popular to be applied to teach students.

Thirdly, on (AI) driven program educators helpful feedback hand, (AI) can't only help teachers and students to craft courses, that are customized to their needs, but it can also provide feedback to both about the success of the course as a whole. Some schools, especially those with online offerings are using (AI) systems to monitor student progress and to alert professors when there might be an issue with student performance. These kinds of (AI) systems allow students to get the support they need and for professors to find areas where they can improve instruction for students who may be given feedback how to learn subject matter. (AI) programs at thes schools aren't just offering advice on individual courses. However, some are working to develop (AI) systems that can help students to choose majors based on areas where they succeed.

Fourthly, on (AI) changing the role of teachers hand, there will always be a role for teachers in education, but what role is and that it entails may change, due to new technology in the form of intelligent computing systems. As (AI) can take over tasks like grade, can help students improve learning, any may even be a substitute for real world tutoring. (AI) systems could be programmed to provide expertise, serving for students to ask questions, and find information or could even potentially replace the teachers for basic course materials. In most cases, however, (AI) will shift the role of the teacher to that of facilitator. Teachers will supplement (AI) lessons, assist students who and provide human interaction and hands-on experiences for students. Hence, (AI) technology has already changed in the classroom teaching method, specially in schools that are online lessons.

Fifthly, on (AI) trial-and-error learning hand, trial and error is a critical pasrt of learning, but for many students, the idea of failing, or even not knowing the answer. An intelligent computer system, designed to help students to learn, to deal with trial and error. Artificial intelligence could offer studrnts a way to experiment and learn in a relatively judgement, free environment especially when (AI) tutors can offer solutions for improvement. In fact, (AI) is the perfect format for supporting this kind of learning as (AI) systems themselves often learn by a trial and error method.

Sixthly, on changing student individual learning hand, (AI) has the potential to change where students learns, who teaches them, and how they acquire base skills, using (AI) systems, software and support. Students can learn from anywhere in the world at any time and with these kinds of programs taking the place of certain types of classroom instruction. So, (AI) robots may replace teachers in some instances ( for better or worse). Educatonal programs is powered by (AI). They are already helping students to learn basic skills, but as these programs grow. Consequently, as (AI) technology can give above advantages to educational organizations. So, in the future, it seems that (AI) can be popular to be used in online or non-online classrooms to assist teacher individual job to raise educational quality.

● Psychological resarch (AI) eduational
social robots and students relationship

Whether can (AI) educational robots bild good social relationsip to students? Some scientists had attempted to do experiments to prove whether (AI) educational robots can do good or bad social relationshp to students.

For example, Knox, W.B. el. (2012) had ever attempted to do two experiments to research whether (AI) robot educational robot can build better or worse social relationship to compare human robot. Their two experiments aim to ask how differing conditions affect a human teacher's feedback frequency and the computational agent's learned performance. The first experiment considers the impact of a self-perceived teaching role in contrast to believing one is critiquing record. The second considers whether a human trainer will give more frequent feedback if the agent acts less ( i.e. choosing actions believed to be worse). When the trainer's recent feedback frequency decreases. From the results of those experiments, they draw three main conclusions that inform the design of agents. More broadly, these two studies indicate as early examples of a nascent technique of using agents as highly specifiable social entities in experiments on human behavior. Thus, it implies (AI) educational robots have ability to learn human teacher to teach students in good social relationship learning environment with students. Even, (AI) educational robots can build better learning relationship to compare human teachers between students, it seems (AI) robots have attractive ability t raise student individual learning interest, after which can applied to assist teachers to teach whose students in classrooms or lecture halls or online classroom channels.

Other scientist had ever attempted to do experiments about " reinforcement learning" (RL) to research how result of interactive supervisory input between human teacher and both robot and software agents relationship. M. Mataric (1997) attempted to do one experiment concerns that reinforcement learning is designed for interactive supervisory input from a human teacher, several works in both robot and software agents have adapted it for human input by letting a human trainer control the reward signal. He aimed to examine the assumption, namely that the human-given reward is compatible with the traditionanl RL reward signal. He described an experimental platform with a simulated RL robot and present an analysis of real time human teaching behavior found in a study in which untrained subjects taught the robot to perform a new task. For the experiment, who reported three main observations on how people administer feedback when teaching a robot a task through reinforcement learning : (a) they use the reward channl not only for feedback, but also for future directed guideance, (b) they have a positive bias to their feedback, possibly using the signal as a motivational channel, and (c) they change their behavior as they develop a mental model of the robotic learner.

Thus, whose experiment concluded that machine learning shall play a significant role in the development of robotic assistants that operate in human learning environment. ( e.g. homes, schools, hospitals, offices). Considering the difficulty of hard-coding all information needd for the robot to play a long term role in az dynamic world, human users will need to be able to easily teach such robots. However, various works have addressed some of hard problems robots face when learning in the real-word.

In robots learning process, what difficulties which will face, the scientist indicated this question concerns how to influence robot's learning abilities: Has RL certain desirable qualifties, such as the learning abilities possibility to explore and learn from unsupervised experience? Many also queation RL as a variable technique for learning in complex real-world environments because of practical problems, such as long training time requirements, non-scaling state representations, sparse rewards ( resulting in slow utility propagation) and safe exploration strategies. As a result, reinforcement learning has been utilized for teaching robots and game characters, incorporating real-time human feedback by having a person supply reward and/or punishment as an additional input to the reward function.

Consequently, the scientist discovered human's teaching method is the most important factor to influence the robot's learning abilities amonf all other environment factors. He also assumed and argued that reinforcement based learning approaches should be reformulated to move effectively incorporate a human teacher. To do this properly, the educational robot must understand the human teacher's contribution; how does the human teach? and what does the educational robot try to communicate from a robot learner? The scientist suggested human trainers ought use these methods to educate robot learners to learn more easily. His main findings indicates reward factor can influence the human robot teachers' motives to teach robot learners to learn, due to more reward can encourage the human robot trainers to teach robots to learn their knowledge and skill. He also found that robot users read the behavior of the robot machine learners and adjust the human robot trainer whose training strategies as whose mental model of the different robot human trainer education method changes. Viweing the human input as a traditional RL reward signal does not take advantage of the fact that a teacher adjusts hose training behavior to best suit the robot educational learner.

In addition to the related RL works mentioned above. Every human robot trainer needs to consider the topic of human input for machine learning systems. Personalization agents and adaptive user interfaces are examples of software that learns by observing human behaviors modeling humann preferences or activities. It empathizes how human teaches the robot learner through interaction, various works address trainable software and robotic agents, exploring explicit human input: learning classification tasks and navigation tasks via natural language, robots that learn by demonstration or/and software agents that learn or training. It seems how to design the robot machine learning software technology will also influence the robot's learning abilities. Thus, human trainer's software learning machine and how to give reward to encourage the human trainer's teaching behavior to let the robot to learn, these factors will influence the robot's learning ability to be applied to education industry to assist human teacher's teaching job successfully. Because how much knowledge and skill, the robot can learn from the human trainer, how much educational knowledge that the robot can own to prepare to teach any students to learn easily. Hence, the human trainer's knowledge and skill will quality to satisfy its primary, secondary and university students learning need.

● Can robots be teachable agents to students really?

In the future, I believe robots have potential to contribute to education by acting as subordinate learners for students teaching. The benefits that the act of teaching provides for one's own learning have long been recognized, tutoring associated improvements in measures like achievement for the tutor role than the lecturer role. However, scientists had confirmed that human teaching robots has been concerned with learning for the benefit of the robot rather than that of the human.

Intelligent autonomous robots are making their way into an increasing range of application areas: Manufacturing, transportation, surveillance and monitoring, rehabilitation, agriculture, service. Hence, if teaching robots can be confirmed to assist teachers to teach students in any schools. It seems that if can be applied to teach manufacturing workers how to produce any products in manufacturing process; it can be applied to teach drivers to drive cars, or teach pilots to control plane engineering machines to fly to sky. So, it can be applied to teach any occupation learners to learn how to operate any engineering machines to replace any human machine training teachers in possible. Hence, it is possible, future one day, robots can be any mahine operating occupation job trainees' teachers ( tutors) or trainers.

In education industry aspect, for computer subject example, robotic systems have potential to increase studetn involvement and motivation and to improve the effectiveness of learning. Possible ways in which they could do so range from small personal robots used as teaching tools for areas like programming and computational thinking to move science-fictional future scenanios wirh humanoid robots supplement or even replacing teacher role in schools.

Recently " teachable agents" have been used as a tool to help students learn, e.g. s student is tasked with training a simulated computer agent on course material, which can lead to a deeper and more committed understanding on the student's own part. For example, a robotic presence in classrooms may one day help to solve problems of teacher shortage, as learning-by-teaching systems have historically been used to do. And the data about human teaching styles that will be generated by widespread use of teachable robots in classroom settings will be invaluable in developing approaches to help robots learn more effectively from humans, where it is rapid and natural instruction of a robot in a new task.

Consequently, whether future robot role is teacher role better or teacher's assistant role or tutor role better in education industry. It depends on the general student's preferable teaching choice in the school. Such as if the school's students prefer to be applied to be taught by human teachers. Then, robots can be choosed to be the school teachers' teaching assistant role to assist the school's teachers' teaching jobs only. Otherwise, if the school students prefer to be accepted to be taught by robots. Then, robots can be choosed to be the school teachers' role to replace human teachers to teach students in the school. Thus, future robots can be either teacher role or teacher's assistant or tutor role for child age, young age or adult age student learning consumers in any primary or secondary or university in future robot computer educational machine development.

Reference

Chou, C. Chan T., & Lin, C. 2003 Redefining the learning companion. The past, present and future of educational gents, computers & education, v.40 n.3, pp. 255-269. April 2003.

Freeman, R. 2000, What is an intelligent tutoring system? Published in Intelligence 11(3): pp.15-16.

IFR, Statistical Department, World Robotics Survey, 2008.Knox, W.B. Breazeal , C.Stone,.O: Learning from feedback on actions part and intended : In proceedings of 7th ACM/ZEEE International conference on human-robot interaction, late- breaking reports session ( HRI 2012).

M. Mataric, " Reinforcement learning in the multi- robot domain," Autonomous robots, vol.4, no 1, pp.73-83. 1997.

Wenger, E. 1987. Artificial intelligence and tutoring systems. Los altos, CA: Motgan Kaufmann.

CHAPTER THREE

# AI human behavioral learning stage

## 3.1Artificial intelligence and the future of defense

Nowadays, artificial intelligence (AI) is widely knowledge to be one kind of the dramatic technology. However, it is expected to continue, to have a disruptive impact on human's private and public life, so defense and security will be no exception. But how exactly will these be affected ? How will (AI) defense and security is incremental in nature?

To research why artificial intelligence (AI) has possible to be used to cause autonomous weapons by human. We need to understand these three aspects of relationship. They include cybersecurity and artificial intelligence and machine learning and autonomous weapon systems relationship between of them.

Firstly, we need to know what is the mean of artificial intelligence and cyber defense/offense? It means defense of critical networks: real time, pattern finding, anomaly seeking, it must utilize machine (AI) learning algorithms to efficiently, and instantaneously respond to potential network threats as well as it means human on or out of the loop. On the loop : it means anomaly detection: human notified, IT analysis, response. Out of the loop: it means anomaly detection: (AI) decides best method of response: quarantine, honey pot monitoring, hack-back. Thus, it is possible that (AI) can be used , such as autonomous cyber weapon.

What is artificial intelligence and autonomous weapons? Autonomous weapons mean one kind of weapon that can be selected and engaged a target, without intervention by a human operator. Are these machines artificially intelligent? I believe the answer is not, because present weapons systems are not capable of human level reasoning. But, (AI) algorithms are presently employed to process sensor data, monitor system health, take and respond to vocal commands manage data, navigate. This, future autonomous weapons systems will require stronger (AI) to be secure and operationally and cost effective. Moreover, self-aware autonomous cyber systems are crucial.

What is cybersecurity mean? It means the ability to control access to networked systems and the information they contain. It is acted to prevent , detect, recover, react. It is application objects concern people, process, technology and it's application goals are confidentiality, integrity and popular availability. Thus, what is cyber weapon mean? Walware means viruses, Trojans, zero-days, worms ransomware, spyware etc. Does it require a particular objective? E.g. military paramilitary or intelligence. Does it require physical harm? E.g. functional harm or interruption? Mental harm? Is (AI) a technological weapon that it is an object or tool? What about when it is an weapon agent?

In simplicity, (AI) can be one of scientific weapons platform. When one day, it is invented to be applied to control war planes to fly to any countries to attack enemies or it is invented to be seemed to human to replace soldiers to bring guns or any weapons go to other countries to attack. So, it is possible that future any war defense planes, (AI) technological automatic control weapon can be replaced of human soldiers or war plane pilots to control any war defense planes to go to different enemy countries to attack them easily. It is very horror matter to threaten global human's ourselves life in the future , if (AI) automatic control war defense planes or (AI) automatic control machine soldiers were invented successfully.

Hence , when (AI) can be applied to weapons platforms, it structures that launch weapons, i.e. jets, ships, vehicles. (AI) platform and weapon and software architecture components are be done one (AI) technological weapons systems. Thus, human will encounter any (AI) benefits or risks ( threats) causes in the same time as soon as possible. If we can predict when (AI) weapon system will be manufactured or invented successfully. Then, we can reduce (AI) weapon systems risks , if we can threaten any (AI) scientists continue to invent any undiscovered (AI) weapons in any time to avoid the future first time (AI) weapon war occurrence in possible.

The (AI) weapon system risk means autonomy: the ability to problem solve technological war , when (AI) weapon system is manufactured successfully, the power to act, how to damage the (AI) weapon system. The power to chance to stop (AI) weapon system manufacturing processes, ability to create a new goals, how to change the (AI) weapon system inventors' or scientists' minds to avoid to apply (AI) tools to achieve attack goals to change to another positive goal. Due to human can't know a prior what an autonomous (AI) weapon system will do.

Although, human is known what (AI) is , but human is also known when (AI) scientists whose emergent behaviors will do to change to do any negative behaviors from positive behaviors. Whatever (AI) weapon system design we use, there will be cybersecurity, problems arising from computation design/complexity. Due to any one (AI) scientist can manipulate the system to act against itself, or who can utilize traditional " cyber weapons" against the (AI) weapon system, or who can manipulate the system to lie to humans, but also due to complexity, there is no way to know if it is lying or not or bounded rationality : satisficing.

Finally, the most serious (AI) technological invention risks are human is unknown these aspects of (AI) absolutely: They are not simple automatic systems, learning reasoning, communication of " self-aware" systems. Thus, human will face (AI) technological invention risks or threats. We need to find any methods to avoid (AI) weapon system is manufactured successfully to avoid (AI) technological war can occur in future anyone day.

3.2 AI system immoral intention

Why (AI) system can be invented to damage our society ? IS it possible to achieve this (AI) damage system successfully? ON (AI) attribution hand, it can be applied to cars, aircraft, which are subject to regulation designed to protect the public from harm and ensure fairness in economic competition. Thus, (AI) safety issue is important to scientists to consider.

IN general, the approach to regulation of (AI)-enabled products protect public safety issue should be informed by assessment of the aspects of risk that the addition of (AI) way reduce any respects of risk that it may increase. Also, where regulatory responses to the addition of (AI) threaten to increase the cost of compliance, or slow the development or adoption of beneficial innovations, policymakers should consider how those responses could be adjusted to lower costs and barriers to innovation without adversely impacting safety or market fairness.

For example, regulatory challenges that (AI) enabled present are found in the cases of automated vehicles. (AI)s, such as self-driving cars and (AI)-equipped unmanned aircraft systems. IN the long run, self-driving cars will likely save many lives by reducing driver error and increasing personal mobility, it will offer many economic benefits. Thus, public safety must be protected as these technologies are tested and begin to mature. Creating safe spaces and test beds for experimentation , and working with industry and civil society to evolve performance based regulations that will enable more uses as evidence of safe operation accumulates. Thus, it implies that any scientists can also invent (AI) system to control weapon defense planes or (AI) automatic machine human to do any soldier's behaviors to attack to any countries easily, instead of none driver automatic control vehicle invention. Thus, (AI) system can be applied to harm to human or achieve to damage our society aim by ourselves in possible.

The rapid growth of (AI) has dramatically increased the need for people with relevant skills to support and advance the field. AN (AI) –enables would demand a data literate citizenry that is able to read, use, interpret and communicate about data and participate in policy debates about matters affected by (AI). Thus, if (AI) technology is applied to assist human's social development and raising life enjoyment or benefits. It will bring positive impact to influence human's future life. Otherwise, if (AI) technology is unsafe to be applied to threaten human's society. It will bring negative impact to influence human's future life. Thus, (AI) scientists need to consider how to apply (AI) technology.

As (AI) technologies move toward deployment, technical expects, policy analysts and ethicists have raised concerns about unintended, consequences of adoption. Use one (AI) to make consequential decisions about people, often replacing decisions made by human –driven bureaucratic processes, leads to concerns about how to ensure justice, fairness, and accountability, the same concerns of human's safety issue. Thus, )AI) expects have cautioned that there are challenges in trying to understand and predict the behaviors of advanced (AI) systems.

Use of (AI) to control physical-world equipment leads to concerns about safety, especially as systems are exposed to the full complexity of human environment. A major challenge in (AI) safety is building systems that can safety transition from the closed world of the laboratory into the outside open world, when unpredictable things can happen. Adapting to unforeseen situations are difficult necessary for safe operation. Experience in building other types of safety artificial systems and, such as aircraft, power plants, bridges and vehicles has much to teach (AI) practitioners about verification and validation, how to build a safety case for a technology, how to manage risks, and how to communicate with stakeholders about risk. The risk means the harm of human's safety of (AI) damage system control machine invention. Thus, any (AI) scientists need consider moral responsibility when who decide to invent what kind of (AI) system machine to aim to bring human's benefits or attribute to human's welfare intention.

Thus, (AI) products safe invention matter will need any scientists' considerations. Because , if (AI) any products are unsafe or harm human's invention in the manufacturing process, it will bring any human's life danger when the (AI) system damage tools are invented successfully and are provided weapons to humans to use to attack other countries easily. It will cause future global human (AI) technological war occurrence.

I shall recommend the solution is necessary of ethical training for (AI) practitioners and students. Ideally, every student learning (AI) , computer science, or data science would be exposed to curriculum and discussion on related ethics and security topics. However, ethics alone is not sufficient. Ethics can help practitioners understand their responsibilities to all stakeholders, but ethical training should be methods for deciding good intentions into practice by doing the technical work needed to prevent unacceptable or immoral (AI) invention outcomes.

Hence, global human needs to concern (AI) weapon system invention security issue. Nowadays, (AI) has important application is increasing role for both defensive and offensive cyber measures. Currently, designing and operating secure systems requires significant time and attention from experts.

Challenges issues are raised by the potential use of (AI) in weapon systems. The United States has incorporated autonomy in certain weapon systems for decades, allowing for greater precision in the use of weapons and safer, more humane military operations. Nonetheless, direct human control of weapon systems involves some risks and can raise legal and ethical questions concern (AI) manufacturing process intention.

The key to incorporating autonomous and semi-autonomous weapon system into American defense planning is to ensure that U.S. Government entities are always acting in accordance with international humanitarian law, taking appropriate steps to control , to develop standards related to the development and use of such weapon systems. The United States has activity participated in ongoing international discussion on Lethal autonomous weapon systems and anticipates continued robust international discussion of those potential weapons systems. Thus, (AI) scientists have responsibilities to manage the potential to be a major driver of economic growth and social progress only, their (AI) intentions are not the global dominance aims absolutely, if (AI) product industry , civil society, government and the public work together to support (AI) positive development of the technology with thoughtful attention to its potential and to managing its invention threat risks to avoid (AI) products to manufacture to be used weapon tools.

Finally, I recommend that as the technology of (AI) continues to develop, practitioners must ensure that (AI) enables systems are governable, that what their inventions need to be openness to let public to know clearly and understandable; that they can work effectively with people and that their operation will remain consistent with human values and aspirations. Researchers and practitioners have increased their attention to these challenges , and should continue to focus on their future any (AI) inventions.

Hence, (AI) safe system ought to be applied to solve the biggest challenges that society faces, such as mobility for the elderly and those with disabilities, smart buildings may save energy and reduce carbon emissions, precision medicine may extend life and increase quality of life, smarter government may solve citizens more quickly and precisely., better protect those at any immoral invention risk and save money.

Moreover, (AI) enhanced education may help teachers give every child on education that opens doors to a secure and fulfilling life. Thus, these are the future human's potential benefits if the (AI) technology is developed to its benefits and scientists ought avoid to manufacture (AI) tools to cause weapon risks and challenges.

Consequently, the main point is that how experts invent (AI) systems. (AI) systems ought not be advanced weapon systems, it doesn't seem to be thought similar human soldiers mind and behaviors. (AI) system ought be systems that think like humans. (e.g. cognitive architectures and neural networks), systems that act like humans ( e.g. pass the test via natural language process, knowledge representation, automated reasoning, and learning), systems that think rationally , e.g. logic solvers, inference and optimization and systems that act rationally e.g. intelligence software agents and embodies robots that achieve goals via perception, planning reasoning, learning , communicating, decision-making and acting function.

In conclusion, it is horror (AI) scientists will invent (AI) systems to be owned human's (soldier's) mind and attack strategic behavior to attack other countries easily, who must need to consider (AI) system ought be invented to own scientists' creating mind and non manual assistance functions for positive attribution to human's society. I expect that (AI) system can only be invented to create human's welfare in our future.

3.3(AI) soldier weapon ethical, social and economic negative impact

In the future, how human can avoid (AI) technological ethical, social and economic negative impact. Scientists need to concern these questions: how to develop of a good (AI) society, how the role and responsibility of the government, the private sector, and the reserch community( including education), in pursuing such a development, whether how the recommendation to support , such a (AI) system development may be in need of improvement.

However, none appers to deliver a comprehensive explicit vision of the role that (AI) system should play in mature information societies. Thus, (AI) 's potential contribution to social good shoud include an in-depth plan for linking in a comprehensive socio-political design questions of responsibility of the different stakeholders, of cooperation between them and of sharable values to understand of a good (AI) positive impact society, not a bad (AI) negative impact society.

Thus, the notion of mature information societies is introduced to stree the importance of addressing the current ethical challenges that (AI) poses in a comprehensive fashion.

It seems (AI) wil invention will be human's moral societal consideration issue. It concerns our (AI) scientists' moral issue, how who invent (AI) system to apply to which kind aspects. IF (AI) system was one direction on war weapon tools to similar to soldier's personal mind or attacking behavior. Then, it will bring poor social safety and poor economy growth our world, due to (AI) scientists' moral is low level.

Thus, the developed country US (AI) technological leader needs to focuse on the impacts of (AI)-driven customatin on the US job market and economy. It represents three specific policy responses to the perceived impact of (AI) on the US economy. They include these three aspects such as: How to invest in and develop (AI) for its many benefits, how to educate and train Americans for the jobs of the future and how to aid workers in the transition and empower workers to ensure broadly shared growth.

The future of (AI) influenced cyber conflicts need more than just the application of current and past solutions in order to ensure security and stability of societies, and avoid risks of escalation. To achieve this end, efforts to regulate cyber conflicts require an in-depth understanding of this new phenomenon, identify the changes brought about by cyber conflicts and the information revoluation, and defines a set of shared values that will guide the stakeholders operating to avoid the international (AI) war occurrence. This becomes clear when considering for example, cyber deterrence. Deploying conventional (cold war) strategies to deter (AI)-influenced cyber conflicts proves highly problematic and the urgent need to foster and coordinate new solutions able to account for the any kinds of conflicts of the cyber demain and of mature information societies to avoid (AI) technological war occurrence in the future.

We hope that in the on-going international conversations and reviews, the US government with further specify how " (AI) system invention law" fit into their vision of the future of society in this case the future of (AI) technological war and conflicts. Hence, (AI) scientists need to concern ethical issues related to (AI), like fairness, accountability and social justice can be addressed through increasing needs. Such as: how the creation of a new body focused on robotics and related (AI) system development to avoid to intent to apply weapon tools to provide advice on the policy, legl and consumer protection issues arising in these fields should be considered.

How to achieve ethical training of (AI) staff and ethical education of the public is certainly important responsibility for (AI) tools ethical behavior and design to the private sector and the citizens : of unique challenges that (AI) brings to society in terms in fairness, social equity and accountability are addresses. Thus, the development of the (AI) technology and defining good (AI) remains problematic. In particular, the US government's innovation driven approach to defining the potential, positive impact of (AI) shows that more could be done to ensure that the opportunities and advantages brought about by (AI) are shared by all society.

An initial on Robotics, based upon the ethical framework and guiding principles is proposed. It should be complementary to legislaton and comprise ethical codes of conduct for Robotics researchers and designers, codes for research ethics committees as well as licenses ( rights and duties) for designers and users. Thus, (AI) robotics invention of safety issues is very important considertion to any (AI) inventions or researchers. Every country's government ought have legal guiding to control their robotics' manufacturing intention. If their robotics (AI) is applied to seem to be soldiers to attack other countries to threaten their people's safety. Then, those (AI) inventors or researchers need to be punished by law.

In conclusion, I believe (AI) technology will be applied to weapon, when it's technological development is nearly mature to able to learn human's mind to do any behavior. During (AI) technology reachs thie mature stage, I predict the (AI) weapon tool , e.g. (AI) soldiers will have chance to be caused. This (AI) invention mature stage has these characteristics such as:

When (AI) invetion reachs this mature stage, computers and robots will develop conscious, intelligent, personified minds. Further, information technology devices and (AI) systems will be implanted into humans, enhancing, psychological and behavioral abilities and allowing for direct communication with artificial intelligent minds. There will be both artificial intelligence (AI) and intelligence amplification (AI) in the relatively near future stage.

During the (AI) invention reachs this mature stage, these will be an ongoing mulit-faceted integration of information technologies and human life. Humans and information technology will cooperate. Humans will increasingly immerse their lives and minds in (AI) systems of technological intelligence and virtual reality. The distinction between humanity and technology will increasingly close dependence.

During the (AI) invention mature stage reachs that the environment will be infused with information technology, becoming animated, communicative and more intelligent. The destinction between the artificial and the natural will increasing close dependence.

During the (AI) invention mature stage will expand through virtual reality, simulated and virtual reality will increasingly into normal reality, e.g. the (AI) weapons is virtual reality to seem to be soldier weapon.

Finally, during the (AI) invention mature stage is as the global expression of the evolving human-technology integration a " world brain" and " world mind" will emerge on the earth. This psychophysical (AI) weapon system will enhance and enrich the capacities of both individual and collective cogniton. This (AI) weapon system is a potential starting point toward the evolution of a cosmic brain and cosmic mind.

Thus, it is possible that the workship raw data was a unique way in which (AI) could be weaponized to cause war, during the (AI) invention stage reachs the invention mature stage. However, (AI) weapon manufacturing factory will be built possibly. In the future, how will we defins and locate (AI) weapon factories. Especially, as these factories are no longer solely buildings , but a mil of virtual and substantially different facilities, particularly as it shifts from a physical assemly and development model to a distributed and flexible network. Needing minimal raw materials to develop (AI) weapons, the phsysical location of their (AI) factories could be anywhere and their identification from the outside, nearly impossible. Given the expanding uses for intelligent and super-intelligent (AI). How will we tell the different form a location that is manufacturing (AI) for the creation of weapons versus creating (AI) for an innovative new gaming platform?

In conclusion, human needs to consider every (AI) scientist's personal ethical or moral mind and research intention and (AI) system invention of (AI) weapon factories cause. During (AI) invention reachs the mature stage if human expects to avoid (AI) technological war occurrence in future one day. The technological development on autonomous military robots, ideally among relevant social groups and actors including human-rights, activists, researchers developers, engineers, philosophers, policy-makers, military authorities, lawyers, journalists and the publis need to consider when human has effort to invent autonomous military robots successfully in the future one day. Finally, some ambitious countries or dominant global countries must like to apply (AI) autonomous military robots to be machine soldiers more than human soldiers if (AI) technology had reached the mature stage. So, future (AI) autonomous military robots will be the next choice of weapon to follow nuclear weapon. If civilians were used as a human (AI) soldiers, the weapon simply ignored them and targeted anyway. This scenario highlighted the dangers of proliferation and quick replication of autonomous weapons. Unlike nuclear weapon, a piece of code for (AI) artificial intelligent soldier could be obtained on the black market and replicated at little cost and the hardware for this type of weapon doesn't require costly or hard to obtain components and materials. Thus, (AI) artificial intelligent soldiers can be manufactured many at cheaper cost. Otherwise, manufacturing one nuclear bomb weapon will spend too much cost. Hence , it is possible that (AI) artificial intelligent soldier will be future new technological weapon to follow nuclear bomb weapon. Hence, any country government needs to legislate to control any (AI) scientists' inventions whether they are attributed benefits or welfares to human or damage human's safety.

- 3.4How does (AI) robots' brain invention influence our lives?

(AI) research modeling the human brain has developed important technologies, and has overcome significant barriers. How will (AI) affect humanity in the near future? How will (AI) change our lives and our societies? Is the evolution of (AI) to humanity, or it represent a threat?

On white collar workers (AI) job replacement aspect, University of Tokyo, Institute of informatics, lecturers who had attempted to do experiments to take (AI) exams over a two year period. The (AI) achieved standard scores of around 50 in each subject, exceeding the norms for humans attempting the tests. The (AI)'s results in subjects emphasizing memorization, such as world history and Japanese history subjects were comparatively high, and the results of the study suggested that an appropriate selection of subjects would give at an 80% chance of passing the entrance exams of 80% of Japan's private universities.

So, if (AI) is applies to human white collar workers' job duties aspect, at this level, if white collar workers were replaced by (AI) in the future, around 30% of current staff would be replaced. Whatever, the outcome, large companies will be represented with two choices. One choice will be to protect their employees, but as a result lose their international competitiveness. The latter choice will enable them to reduce the cost of general duties, financial management procedures, accounting etc. general administrative job duties of cost in offices. Hence, it seems that (AI) will be possible to be invented to own human's brain ability to do some mind jobs in future on day.

However, the method called " deep learning" must be developed to cause (AI) to match human's brain ability as well as these were dramatic advances in technologies, such as image recognition and voice recognition, which form the foundation for (AI). Nowadays, this new method called" deep learning" does not reach the matured and stagnated stage. It needs to wait human to continue to invent to let (AI) to match human brain to achieve 100% owning human's mind ability. Nowadays, (AI) industry product include cleaning robots, smart TVs and future (AI) product development market. It will include self-driving vehicles, drones, and nursing robots.

On (AI) weapons applied aspect, if (AI) can be invented to own human's brain judgement and analytical abilities. Then, it is possible that it can be applied to attack enemy to cause war effect. For example, if weapons such as missiles were equipped with (AI) in the future, they would become able to decide on their own targets. Hence, human needs to apply restrictions when necessary.

On (AI) applied to analyzing information collected technological aspect, nowadays, every one will use wearable terminals to connect to the internet to obtain various types of information as well as computers will collect and analyze information on people. Our lives will probably be more reliant on these internet technologies than they are on smartphones today. When, (AI) can match human brain to own mind ability.

Then, (AI) can be applied to do any analyzing information and collection job duties aspect to raise large information restoring and remembering efficiency. For example, (AI) will be generally used and will be extremely useful in analyzing the information collected from wearable devices and stored in the cloud. (AI) will enable wearable devices to be of real assistance in our lives offering their users more intelligent support.

Rather than allowing (AI) to develop on serves, as something separate from humanity. It will be more meaningful to encourage its development via wearable devices, situating it under the control of human intelligence. The intelligence of (AI) will increase rapidly in the future. If this increase in (AI) occurs under human control, enabling humans to increase their own abilities, then surely it will be possible for us to put up a degree of resistance to the opposite scenario, the domination of (AI) over humanity. Hence, if (AI) can be invented to remember and store and make analytical judgement to collect any information from internet. Then, it will bring the effect, such as large international organizations‘ (AI) internet storage robots can bear in mind factors, such as competitors' privacy or business secret information, such as the loss equality between people and threats to privacy that will be stolen form the owning (AI) storing internet information remembering robots.

Consequently, what is the effect of successful invention of (AI) matching human brain's mind ability? (AI) present computers are adequately able to reproduce the emotional, conceptual and intuitive abilities of humans. Because of this, it is important that we should envision potential future problems that may manifest when we consider how to employ wearable devices. It will be essential to enhance our technologies in order to ensure that we can use (AI) under human control.

However, when a goal has been set. (AI) will implement an appropriate means for its realization. (AI) will be need as a tool by human society. If the capacities of analytical and judgement mind abilities of (AI) brain exceed those of human brain, it is difficult to imagine the type of technological, then singularity is represented by the creation of an (AI) by another (AI).

It is important that we rapidly and accurately predict these developments, when image recognition and other individual technologies are functioning at a high level. There will be a considerable matter in different sectors of (AI) industry development.
Today, however, machines have become able to decide for themselves what they will learn, making it difficult to copy human's mind ability. What we must consider when machines exceed humans and (AI) surpasses human capabilities. May technologies exceed human capabilities, cars are faster than humans, planes are able to fly. Consequently, it brings a question that human needs to consider: When does (AI) brain technology be invented to reach the most reasonable stage to be accepted or stopped by humanity?

- 3.5What is artificial intelligence
human brain invention?

A machine is likely to achieve the ability of a human brain. Does it a scientific story? Some scientists has predicted that a US$1,000 personal computer will match the computing speed and capacity of the human brain by around the year 2020 year. With human reverse engineering, human should have the software insights before 2030 year. it is possible that of machine intelligence and exotic new technology for faster and more powerful computational machines from cellular automata and DNA playing cheese game competition case example, it proves that (AI) had been invented to own human's analytical and judgement ability to exceed the best cheese game human player's brain analytical and judgement ability. Then, it seems that (AI) will have possible to be built machine brains to achieve the exceed level of human brain's analytical and judgement ability in the future one day.

Supposing we scan someone's brain and restate the resulting " mind file" into suitable computing medium. Will the entity that emerges from such an operation be conscious? How have advances in electronic communications changes power relationship? For electronic book publishing case example, a book that looks at the principles companies must adopt to meet the needs and desires of this new kind of client. So, such as paper book can be changed to electronic book for human to read. Why can't human brain be changed to (AI) machine brain to do human's analytical mind and behavioral mind of activities to replace to do any human's daily analytical and behavioral mind activities?

Over the next few decades, machine achieve super intelligence, human will encounter a dramatic phase. Will it be a "WALL" a barrier as conceptually the event of a black hole in space. Such as (AI) brain invention case, an " AI singularity" ruled super-intelligence AIs, or a gentler " surge" into a post human era of agelessness and super-intelligence brain. Will future technology, such as bio-engineered pathogens, self replicating nan robots, and super smart robots run and accelerate out of control, perhaps threatening the human race?

If one day, (AI) brain is invented to achieve agelessness possibility. It means human's brain will be old to lose mind and analytical ability when human's age is increasing. Otherwise, (AI) machine brain age won't lose mind and analytical ability, due to (AI) machine is no age increasing possibility. It is a machine brain. If (AI) machine brain can be built successfully. Scientists need to consider technological ethic matter, such as the challenge of guiding nanotechnology in a constructive direction, advances in nanotechnology and related advanced technologies can not be inevitable, any broad attempt to relinquish nanotechnology would interfere with the benefits. When actually making the dangers worse.

Keeping in mind that intelligence machines are already making their way into our blood stream. There are dozens of projects underway to create blood-stream based " biological micro electronic- system" (bio MES) with a wide range of diagnostic and therapeutic applications BioMEMS devices are being designed to intelligently pathogens and deliver medications in very precise ways. For example, a researcher at the University of Illinois at Chicago has created a ting capsule with pores measuring only seven nanometers. The pores let insulin out in a controlled manner, but prevent antibodies from invading the pancreatic Islet cells inside the capsule. These nano- engineered devices have cured rated with type I diabetes, and there is no reason that the same methodology would fail to work in humans. Similar systems could precisely deliver dopamine to the brain patients, provide blood-clotting factors for patients with hemophilia and deliver cancer drugs directly to tumor sites. A new design provides up to 20 substance-containing reservoirs that can release their cargo at programmed times and locations in the body.

Another brain health technological related invention case, such as Kensall Wise, a professor of electrical engineering at the University of Michigan, who has developed a tiny neural probe that can provide precise monitoring of the electrical activity of patients with neural disease. Future designs are expected to also deliver drugs to precise locations in the brain. Also, kazushi Ishiyama at Tohoku University in Japan has developed micro machines that use microscopic-cancer tumors.

A particularly innovative micro machine developed by Sandia National labs has actual micro teach with a jaw that opens and closes to trap individual cells and then implant them with substances, such as DNA, proteins or drugs. There are already at least four major scientific conferences on bio MES and other approaches to developing micro-and nano-scale machines to go into the body and bloodstream. All these inventions are related to how to apply machines to copy human's brain knowledge in order to achieve to do any human's brain functions.

Finally, for Freitas envisions micron-sized artificial platelets invention case example, who could achieve hemostasis ( bleeding control) up to 1,000 times faster than biological platelets. Freitas describes nano-robotic microbivores ( white blood cell replacement) that will download software to destroy specific infections hundreds of time faster than antibiotics, and that will be effective against all bacterial, and fungal infections with no limitations of drug resistance.

Consequently, such as above machine health scientific invention cases, there were many scientists had invented any health machines to apply drugs to transfer to human's brain to attempt to reduce human's disease causing risks, such as reducing cancer cell increasing number. Why it is no possible that scientists can attempt to invent (AI) brain which can own human's mind ability to judge or analyze any matters to give opinions in order to exceed human's judgement and analytical ability.

### 3.6How can artificial intelligent brain satisfy to human beneficial and natural needs?

Nowadays, new scientists' most familiar form of this vision in our times is genetic engineering. Specifically, the prospect of designing better human beings by improving their biological systems of a small, serious and accomplished group of tailors in the field of artificial intelligence and robotics. Their goal is a simply new age of post-biological life, a world of intelligence without bodies, immortal identity without the limitations of disease, death and unfulfilled desire. If human can understand why this fate is presented as both necessary and desirable, human might understand modern science can help us to enter the good life and good society stage when (AI) brain is invented by scientists successfully in our future life.

How can (AI) beneficial brain satisfy to human natural need? For relatively recent example, similarly as a long term trend beginning with the first mechanical calculators, the evaluation of computing capacity increases in speed over time and decrease in cost. From biological evolution has been invented to influence human brain, an electronic chemical machine with a great, but finite number of computer neuron connections, the product of which we call mind or consciousness. As an electro-chemical machine, the brain obeys the laws of physics, all of its functions can be understood and duplicated. And since computers already operate at far faster speeds then the brain, they soon will rival or surpass the brain in their capacity to store and process information. When happens, the computer will at the vary least, be capable of responding to stimuli in ways that are indistinguishable for human responses. At that point, we would be justified in calling the machine intelligent, we would have the same evidence to call it conscious that human now have when giving such a label to any consciousness other than our own.

Image

At the same time, the study of human brain will allow us to duplicate its functions in machine circuitry. Advances in brain imaging will allow us to " map out" brain functions, allowing individual minds to be duplicated in some combination of hardware and software. The result, will be a world that is remade and reconstructed at the atomic level through nanotechnology, a world whose organization will be shaped by an intelligence that surpasses all human comprehension.

Whether or not today's humans are willing or able to " download" their brains into machines, there will come a time when all human beings will be intelligent machines in the future. Computer hardware will continue to get faster, cheaper and more powerful computer software will increase in sophistication. Brain research will continue to explore the " mechanics" of consciousness. Nanotechnology will continue to develop.

There are powerful incentives, commercial, military, medical and intellectual that will drive many of the advances that the extinctions desire if for very different reasons. Much of the work in artificial intelligence and robotics is open to the same defense that is made on behalf of biotechnology: If we don't do it, they will and why suffer or be unhappy when some new agent or invention is available that will solve the problem.

Finally, we already accept significant artificial argumentation and replacement of natural body parts when those parts are missing or defective. Over time indistinguishable from or " superior" to their biological counterparts as they employ increasing computer processing power. There are powerful incentives, commercial, military, medical and intellectual that will drive many of the advances that the extinctions desire, if for very different reasons. Much of the work in artificial intelligence and robotics is open to the same defense that is made on behalf of biotechnological if we don't do it, they will and why suffer or be unhappy, when some new agent or invention is available. That will cure the problem.

Finally, we already accept significant artificial augmentation and replacement of natural body parts when those parts are missing or deductive. Over time, such replacements are only likely to get more useful and perhaps eventually indistinguishable from or superior to their biological counterparts, as they employ increasing computer processing power. Nor is there an obvious distinction between using manufactured chemicals to fight disease and using " smart" nanotechnology. The extinction project is begun by offering new routes to fulfilling old promises about doing good for human beings. But, it doesn't necessary end.

In connection with machine intelligence, it does not seem very promising to try to limit the power or ability of computers. The danger ( or promise) that computers might develop characteristics that lead some people to call them conscious and that this age of intelligent machines would mean our extinction seems remote when compared with their practical benefits. We already rely so heavily on computers that the incentives to make them easier to use and more powerful are very great. Computers already do a great many things better than we can, and there seems to be no natural place to enforce a stopping point to further abilities. Certainly mechanistic and reductionist assumptions about society, ethics and psychology the notion that we are atoms or animals, driven by chance or instinct, run deep in the present world.

3.7Artificial intelligent brain future
innovation and attribution

- (AI) brain invention successful factors

(AI) brain will bring what attribution to influence human's positive impact. How (AI) brain will be invented to apply to any businesses' needs. By how much the (AI) brains might exceed us remain unknown, but it could potentially be by a very significant degree. Future (AI) brain invention will have noted similar growth in everything from hard-drive storage density to the price and speed of DNA sequencing. On key feature of this technological growth that has not been adequately measured in the degree to which technology is becoming more intelligent.

3.8 (AI) brain test experiment

When there is an intuitive sense that (AI) programs/brains today are more capable than those of age, and that those were considerably " smarter" than the serial instructions that passed through the first supercomputers. Scientists will carry on testing (AI) to do any experiments to improve (AI) brain development. They will assess the progress of artificial (general) intelligence, but the need for intelligence tests and tests for other cognitive abilities will be tested in the forthcoming decades for bots, robots, avators, " animats" etc. and any collective system of these and biological systems ( humans and non-human animals).

When (AI) brain technology is invented successfully, the idea of a super intelligent computer means invention successfully also. Whether super intelligent computer or (AI) brain can be invented successfully. Similarly, many think that the future beyond the technological singularity is unknowable or even unimaginable. Since its conception, the idea has been examined and explored by technologists. (AI) brain invention will be mean human-equivalent (AI) vs human-level(AI). Many machine intelligence tests is the exclusive focus on identifying systems that achieve human equivalence. Considering our experience with studying non-human animal intelligence.

The need to distinguish human equivalent (AI) from human level (AI) seems critical. Perfect human intelligence , such as (AI) brain invention is likely to be extremely difficult to achieve. Potentially every aspect of the biological processes involved would need to be translated with very high fidelity.

Human level (AI), such as (AI) brain invention is another matter. Achieving capabilities that are equivalent to those of the human mind could be very feasible if it is not limited to perfectly processes involves. For instance, some pattern recognition algorithms are already superior to human abilities. This machine ability is not achieved by duplicating the processes our brains use, through some of the methods have been inspired by them. If researchers had been limited to replicating the brain's mind processes, We would still be waiting for the development of a (AI) brain machine equivalent.

Tests that would seen to have a reasonable chance of successfully testing non-human intelligence are those that use mathematics to define the value of a given challenges for (AI) brain invention. Such as Capability-test has some potential to generate meaningful data about (AI) machine intelligence, e.g. others in complexity theory test, it presents a series of abduction and prediction problems, similar to those in standard IQ tests to (AI) brain. Hence, IQ test and C-test will be potential tests to test (AI) brain ability.

(AI) brain test goals include that to identify a range of possible types of mind such as: super-fast human mind, mind with operational access to its source code, any mind capable of general intelligence and self awareness, general intelligence without self-awareness, self-awareness without general intelligence, super-logic , machine without emotion, mind capable of imaging greater mind or creating greater mind to compare human's brain abilities. Thus, any IQ or Capability test experiments aim to evaluate whether (AI) brain mind ability which can exceed to human brain mind ability. Thus, if future one day, scientists could prove (AI) brain mind ability can exceed to human brain mind ability. Then, they believe (AI) brain invention has ensured to achieve success.

3.8.1(AI) brain invention successful factors

Hence, scientists want to invent (AI) brain successfully. They need to solve these challenges. Such as How robots and computers have progressively supplemented humans, initially only in relatively simple computational and manipulation tasks, but more recently in higher cognitive tasks that used to be the pre negative of the human brain, including language, mathematics, probabilistic reasoning and decision making.

An important challenge is how to enhance the productive interactions between humans and artificial intelligence? The important challenge include these major successful factors to invent (AI) brain technology, such as:

What is the state of the art in (AI) software and machine learning?

Can all aspects of brain function be manufactured by artificial system?

What is the proper form of mathematics that may capture the operation of minds and brains?

How to make (AI) brain feels consciousness?

What would it be taken for a machine to pose a sense of art in (AI) brain software and (AI) brain machine learning by artificial system?

Will machines soon surpass us in all domains of human competence?

What is the proper form of mathematic that may capture the operation of minds and brains?

What is consciousness to (AI) brain invention?

Could a machine be endowed with an artificial consciousness?

What would it take for a machine to posses a sense of self?
Will intelligence machines soon pose a danger to humanity of (AI) brain is invented to reach the mature stage successfully?
Is it possible to design and construct an intelligent robot with an artificial brain sense of ethics?
How can we enhance the humanitarian uses of artificial intelligence brain and owning mind ability of robotics, in particular in the field of education, health and emergencies?
Consequently, above all these challenges, I recommend scientists need to solve to achieve (AI) brain invention in order to reach (AI) brain invention mature stage easily.

### 3.8.2Artificial intelligence brain invention opportunities and challenges

If scientists focus wrong direction to invent (AI) brain, it will bring wrong marketing development to attribute any benefits to human. So, they need to reduce a mismatch of timescales between the pace of commercial innovation and (AI) brain invention process, reduce an underappreciation of the fundamental unpredictability of (AI) brain autonomous systems and reduce a lack of a university agreed upon conceptual framework for any (AI) brain invention and reduce a disconnect between the (AI) brain design of any kind of (AI) autonomous robots. Thus, scientists need examine these gapes, provide a roadmap of opportunities and challenges and identify areas of any (AI) beneficial functions to be attributed to human to use for our daily life needs.
Future (AI) brain market opportunities, it is rapidly growing innovations in digital -electronic and information technology had development of new intelligence, surveillance, and reconnaissance platform and battle management capabilities, precision-strike weapons, stealth aircraft, smart weapons and sensors and tactical exploitation of space ( e.g. GPS).
Any one of these aspects will be (AI) brain future marketing development opportunities. Future (AI) brain invention of so called " narrow AI", (i.e. non-sentient artificial intelligence, whose problem-solving capability is confined to one narrow task. For example, (AI) brain needs to find the best method to win any one of chess player in any both human and (AI) robot chess playing game.
In smart weapon strategy industry, (AI) brain is needed to design how to analyze or mind to protect whose country to avoid enemy attack in any war

by the best weapon protection strategy. In education industry, (AI) brain is needed to design how to analyze or mind how to assist teachers to educate whose students by the best education method. In aircraft industry, how to design (AI) brain to analyze or mind to assist pilot to make the most correct flying direction judgement to fly in the most safe way. In space exploitation industry, (AI) brain is needed to design how to find undiscovered natural resources to supply to human to use in anywhere space.

Thus, future (AI) brain invention needs have these features/characteristics to be designed. They include: (AI) learned on its own, where to find the information it needs to accomplish a specific task, (AI) can predict the immediate future from studying any matter, (AI) automatically needs to be inferred the rules that govern the behavior of individual robots within a robotic swarm simply by watching, (AI) needs to be learned how to navigation the acquired memories and experiences, much like a human brain, (AI) speech recognition needs to be reach human parity in conversational speech, (AI) communication system needs to be invented its own encryption scheme, without being taught specific cryptographic algorithms ( and without revealing to researchers how its method works, (AI) translation algorithm needs to be invented to remember fluent language to more effectively translate between any two languages ( without being taught to do so by humans), (AI) brain system interacted with its environment ( via virtual environment) to learn and solve problems in the same ways that a human child can do, (AI) based medical diagnosis system needs to be achieved 99% percent accuracy in any medical reviewing researches ( at a rate minimum 30 times faster than humans), ( AI) poker playing program brain development needs to be defeated some of the world's best human poker players during a minimum three-week-long tour, (AI) brain development needs to be effectively " read minds" of human test subjects looking at pictures of faces, via functional magnetic reasonable images of brain activity.

Consequently, (AI) future brain development needs to follow above directions to be invented to attribute to human's satisfactory needs.

3.9 (AI) brain legal remembering attribution

Future, (AI) robots can assist lawyers to deal any legal cases more efficient. If human understands that smart (AI) technology is not to replace human lawyers, but to make a better lawyer that forces who to use, emotional intelligence, and capital on lawyers‘ mind, then human lawyer have made the first step in future, proofing whose legal service business from (AI) legal robots' assistance. (AI) promises to be a real advantage for today and tomorrow' lawyers having to deal with the rate of legislative evolution and technological change.

In future, for the better with regard to technology in any law firms. According to the ALM 205 law tech. survey 95% of firm leaders and technologist respondents agreed with recent decisions by management regarding the firm's technology in legal service profession.

How can (AI) robots be applied to legal service industry by legal service firms? The (AI) reality in the legal world, it includes in relation to the four key elements of legal service provision, such as commodity, research, reasoning and judgement, exist and can support or replace certain aspects of every lawyer individual jobs both fee earning processes and business processes can be supported and/or replaced by expert systems, cognitive computing, robotics automated systems, (AI) and the machine learning, clients demand and expect more speedy, accurate, expert, creative, intuitive and accessible legal advice, it can assist young lawyers to innovate / tech. focused firm, reducing pressure both from within law firms ( or in house teams) and from clients to respond to the demand for client-designed service from (AI) robots assistance, technology related projects that are both user and client -centric need to be implemented successfully.

Due to the deployment of (AI) robots in the legal ecosystem where lawyers, firms, general counsel and clients are beginning to believe (AI) capability technologies, (AI) robots can assist lawyers to increasingly become more productive, efficient, accurate, better quality, less labor intensive and time intensive and the role of the lawyer is gradually changing. If we break down a lawyers' tasks in a legal project from beginning that can handle the majority of these four tasks far more quickly and accurately than human lawyer. (AI) robots can handle legal task in the four aspects as below:

First in legal aspect, it can be used for deep research and processing, such as extracting specific pieces of information from land registry documents, (AI) technology is placed top of a document set including client guidelines and similar forms. It searches through documents and extracts key data points to provide a report of data for improved coordination with clients.

Second on managed services technology aspect, (AI) platform which could have a huge advantage for general counsel and law departments in corporations and for clients of all company sizes.

Third on reading aspect, (AI) brain program that reads and analyzes , e.g. clauses in loan agreements. Its program helps its lawyers through transactions and points. Then toward the correct precedents of each stage of a process.

Fourth, on academics aspect, (AI) robots can assess the merits of personal injury cases. It can automatically review high volumes of contract documents to identify provisions that could potentially be impacted by contract law regulation.

Thus, all of these systems can handle large quantities of structured and unstructured data, and assist with the process management, research, and reasoning elements related to legal issues. Thus, in future, (AI) brain development can be invented to apply to knowledge research job, such as legal industry.

Artificial legal intelligence presents a thought-provoking approach to both computational models of legal reasoning and the use of evolutionary thinking about the law. The visions of computerized artificial legal intelligence , a vision of developments in both technology and legal history. A number of creative research projects have applied artificial intelligence techniques to the domain of legal reasoning.

Consequently, ( AI) brain for legal industry marketing development ought concentrate on those three fields of artificial intelligence at most relevant to work in the legal areas in order to achieve the excellent attribution. Such as case-based reasoning, expert systems and neural networks. Artificial intelligence program, such as the legal reasoning programs. Thus, (AI) brain invention needs to own these human legal concept knowledge in order to achieve (AI) legal brain program development successfully.

3.9.1 Whether human mind can create to (AI) brain mind

How can (AI) scientists build a machine that think? In fact, there was general agreement that minds can be existence on non-biological substrates and that algorithms are of central importance to the existence of minds. However, there are much debate about the raw hardware power present in organic brains, such as (AI) brain.

I think (AI) scientists need to invent powerful hand ware, e.g. commercial digital signal processing might be, giving an appearance even to digital operations, but nothing would ever make up the intellectual runaway that is the essence of the singularity.

I also think raw hardware power is not be able to organize the parts to behave in a super-human way as well as I also think powerful software complexity is the main factor to solve (AI) brain mind invention challenge. Hence, future super-human (AI) brain invention will ought consider how to invent superhuman software more than hardware, because software can store any memory, i.e. human mind. When, (AI) brain invention which can achieve to own human mind ability. Then, (AI) scientists need to consider ethic matter: Does the future of (AI) pose an existential threat to humanity? How do we present learning algorithms from morally objectionable biases? Should autonomous (AI) be used to kill in warfare? How should (AI) systems be in our social relations? Is it permissible to fall in love with an (AI) system? What sort of ethical rules should (AI) like a self-driving car use? Can (AI) systems suffer moral harms? All those ethic matters. I think (AI) scientists need to consider after (AI) brain owns human's mind ability because it is possible that (AI) robots will harm human if they are educated to do any wrong or illegal or immoral mind ability. Thus, (AI) scientists need to consider (AI) robot's moral mind and judgement behavior.

### 3.9.1 Brain-inspired intelligent robotics

How to solve fundamental problems in the areas of brain sciences and brain-inspired intelligence technology? One way scientists seek to accomplished the mission is to develop brain-inspired hardware including intelligent devices, chips, robotic systems and brain inspired computing systems. (AI) scientists ultimate goal, but since the robot's memory and learning after the human brain, there is still much to learn about neurobiology before that goal is attached.

In (AI) brain university research aspect, two schools of thought have emerged in robotics: bio-logically robots that include a body, sensor and actuators, and brain-inspired computing robot.

Robots have found increasing applications in industry, service and medicine, due in large part to advances achieved in robotics research over the past decades, such as the ability to accomplish complex manipulations that are essential for automated product assembly. However, robots still have these weaknesses which need to be solved if (AI) scientists expect (AI) brain invention can be success. Robots still lack truly flexible movement, have limited intellectual perception and control, and not yet able to carry out natural interactions with human. These deficits are especially critical in service robots. A critical concern of government, academia, and industry is how to advance research and development for the key technologies that can bring about the next generation of robots. Developing robots with more flexible manipulation, improved learning ability and increased intellectual perception will achieve the main goals for (AI) scientists' solutions.

Consequently future (AI) brain -inspired intelligent robotic invention needs have these competitive or attractive abilities or strengths to compare computer storage ability. Such as (AI) brain needs have perception to exceed computation ability, adaptation exceeds computing speed, flexibility exceeds, computer memory access speed, cognition exceeds computer memory lifetime, learning exceeds computer memory capacity and innovation exceeds computer memory storage abilities. Hence, (AI) brain innovation must need to exceed general computer storage, lifetime, capacity, learning abilities if (AI) scientists expect (AI) robots will be popular to be applied by any service, manufacturing, education industries.

3.10How can web intelligence need (AI) brain informatics?

Brain informatics (BI) invention will be a new inter-disciplinary field that systematically studies the mechanisms of human information processing from both the macro and micro view points by combining experimental cognitive neuroscience with advanced information technology. (BI) studies human brain from the viewpoint of informatics ( i.e. human brain is an information processing system) and uses informatics ( i.e. WI centric information technology) to support brain science study. It seems that (AI) scientists need to further understand how human intelligence and brain sciences development through brain sciences fosters innovative web intelligence research and development because innovative web intelligence research will have ability to assist (AI) scientists to research how to invent (AI) brain robotic technology more easily. The synergy between ( web informatics) (WI) and ( brain informatics) (BI) advances our ways of analyzing and understanding of data, knowledge, intelligence, and wisdom, as well as their interrelationship, organizations and creation processes. Web intelligence is becoming a central field that information technologies and artificial intelligence to achieve human level web intelligence.

(WI) may be viewed as applying results from existing disciplines , e.g. artificial intelligence (AI) and information technology (IT) to a totally new domain the world wide web. (WI) may be considered as an entrancement or an extension of (AI) and (IT), (WI) introduces new problems and challenges to the established disciplines.

Thus, developing human-level web intelligence will be seem to develop brain level artificial intelligent technology. Because brain informatics (BI) is an interdisciplinary field to systematically investigate human information processing mechanisms from both macro and micro points of view by cooperatively using experimental, computational, cognitive neuroscience, and advanced (WI), centric information technology. It attempts to understand human intelligence in depth, towards a holistic view at a long term, global vision to understand the principles and mechanisms of human information processing system (HIPS).

So, I recommend (AI) scientists needs to research how to invent brain informatics technology and web informatics technology to achieve how to apply (AI) brain to analyze and judge or mind web informatics ability to compete internet ( web site) communication tool product industry. Hence, if (AI) brain can be applied to web informatics industry will increase attraction to an internet user market in the future.

3.11Why model of sustainable development environment industry be (AI) robot attractive market

In the world scientific literature various conceptions of sustainable development are found, however, their basis are formed by three dimensions: environmental, economic and social development. The biggest attention is concerned to the environmental dimension which forms the basis of existence of social environment and economy.

The reason is emphasized that each person must preserve and manage natural resources as the basic of economic and social development. However, in the sustainable development strategy, the sustainable development is understood as among environment protection, economic and social society that form the basis to achieve the universal welfare for present and future generations, it is difficult to let human to adapt to live in pollution environment. Thus, environment pollution will be human's concerning issue. It implies how to protect environment pollution which will be human's future challenge. If (AI) brain invention can be applied to how to solve environment pollution challenge. It is very attractive attribution to human (AI) brain environment protection function can be invented to include such as : how to apply (AI) brain to do mind to give opinions or do any environment protection behaviors/activities to assist human how to effective use of natural resources, how to effective use of universal economic society's welfare, do strong social guarantees during the period of strategy's implementation ( until 2020 year to achieve global (AI) robots environment protection mission from (AI) robots' behavioral assistance. Because environment pollution will influence world's climate to be worse to cause our food or vegetable can not grow up easily. Then, human will face food / vegetable shortage challenge. If (AI) brain invention can be applied to solve environment pollution aspects, then human will avoid food / vegetable shortage crisis.

3.12 How to invent (AI) brain's environment protection ability?

All similar problems significantly promotes scientists to research for new technologies of data extraction progressive software of data extraction operating on the basis of artificial neural networks allow finding the relations among various types of data in the huge data. Due to the data extraction technologies, it is possible to prove empiric observations to group, to process and model big amounts of data, distinguishing unknown schemes in data and using them in future activities ( Rotman M. J., 1995).

It seems scientists believe gather every country's climate environment data to predict environment climate change will have chance to avoid environment pollution challenge. In addition, the traditional statistic methods applied to process digital data of the indicators of sustainable development environmental dimension are not able to analyze the present situation of environmental dimension in the context of sustainable development to present possible reasons of change of environmental dimension problems to generate future forecasts characterized by a high level of accuracy.

Consequently, if (AI) scientists can apply (WI) web informatics technology and (BI) brain informatics technology both to apply to (AI) brain technology to gather global climate environment daily change data. Then, I believe (AI) brain invention can assist human to solve future environment pollution challenge Then, it is possible that (AI) brain can attribute to environment protection or how to give opinions or choose to do any environment pollution activities to avoid global warming crisis.

3.13 (AI) human behavioral learning Asia market

When (AI) brain is successful invention, I think Asia will be a new market to need (AI) brain attribution for Asia consumer needs. I believe Asia people expect (AI) can attribute to let them to use. The (AI) ability includes the ability of machines and systems to acquire and apply knowledge, and to carry out intelligent behavior.

This includes a variety of cognitive tasks ( e.g. sensing, processing, oral language, reasoning, learning, making decision) and demonstrating an ability to move and manipulate objects according). So, future Asia people (AI) consumers expect (AI) robots can attribute to whose society's needs, such as: how to apply intelligent systems to use a combination of big data analytics, cloud computing, machine-to-machine communication and the internet of things ( IOT) to operate and learn. Asia people expect (AI) robots can give beneficial attribution to them, e.g. talking or playing a game for Asia young entertainment market; (AI) robots need to reflected by physical substance ( such as any talking or playing game robot player).

In this sense, (AI) is like a human brain. For Asia (AI) robot service industry need, Asia (AI) robot clients expect to use soft robotics ( robotic process automation) can be used to meet Asia (AI) robot service consumers' expectation of automation repetitive tasks and common processor needs, such as client servicing and sales without the need to transform existing IT system maps ( e.g. (AI) salespeople robots, or (AI) service robotic).

In (AI) office task aspect, Asia office consumers expect algorithmic game theory and computational social choice of (AI) robots to be attributed to replace some office human workers' tasks. Such as (AI) systems tat address the economic and social computing dimensions of (AI), such as how systems can handle potentially incentives, including self-interested human participants or firms, and the automated (AI) -based agents representing them, e.g. complex or simple office administrative tasks, e.g. typing, accounting, filing, etc. general office tasks which need human office workers who use computers to work to be replaced by (AI) robots to do. So, Asia office (AI) robots users who expect any office human administration tasks can be replaced by (AI) robots to do.

In Asia computer vision market, Asia computer vision ( image analytics), users expect (AI) robots can be replaced to human computer image workers to shorten time to work, or raise image quality to be more clear in the process of pulling relevant information from an image or sets of images to advanced classification and analysis. Such as hospital or clinic x-ray image vision (AI) robot invention, photo image (AI) robot invention etc. any Asia image industry (AI) robot market need.

In Asia collaborative systems work with human (AI) robot market, Asia autonomous systems robots users who expect (AI) robots can be applied its models and collaborative systems to help them to develop autonomous systems that can work collaboratively with other systems and with humans, e.g. car manufacturing, computer manufacturing or any machine manufacturing products. So, Asia machine related manufacturing product industry manufacturers who expect to apply (AI) robots who can assist factory human manufacturing workers to manufacture any products in the short time efficiently.

In Asia language teaching (AI) robot market, language educators expect (AI) robots own natural language processing ability, algorithms that process human language input and convert it into understanding representation, such as Asia translation education market ( PWC).

Thus, Asia language teaching businessmen expect (AI) brain invention which needs to be designed to own these above abilities of (AI) robot's manufacturers expectation to provide to them to use from any (AI) robots import.

3.14Is the brain a good model
for machine intelligence?

The actions of a human " computer" using paper and pencil to perform a calculation ( as the world meant ), into a formalized machine, manipulating symbols on an infinite paper tape. But I believe it still has challenges to influence human brain can be invented to machine intelligence successfully. I think (AI) scientists need to solve these further challenges to influence (AI) brain invention success. The limitations include such as below:

Computation is based on functions of integers is limited. They must let computer data can change to words, then words can change images, then any images can change to storage to let (AI) brain to remember to do any analytical and judgement able tasks to make any actions in the short time finally. It is (AI) brain behavioral and analytical mind speed limitation.

Moreover, anther limitation concerns biological systems clearly difference, they must respond to varied stimuli over long period of time, those responses any changes alter their environment and subsequent stimuli. The individual behaviors of social insects, for example, are affected by the structure of the home, they build, and the change their behaviors. Nowadays, (AI) brain invention is called computational neuro science, which have assured that the brain is a computer, it means a machine that is algorithms and architectures. Second, neuro science findings may validate the energy ability of existing algorithms being integral parts of a general (AI) system. It means (AI) robots change adapting environment limitation. It means how (AI) brains adapt to choose to make any analytical mind as well as how to be influenced by external environment to do their behaviors or actions in the efficient way, e.g. how to manufacture many cars efficiently in one factory in the short time.

Consequently, to solve these limitations, we need to know how to apply correct conceptual knowledge to let (AI) robots to learn, e.g. for example, if we know how conceptual knowledge was formed from perceptual inputs, it would crucially allow for the meaning of symbols in an artificial language system to be grounded in sensory " reality". When (AI) scientists can achieve how to solve all above limitation challenges, then (AI) brain invention will achieve more easily.

Reference

PWC, Sizing the prize, see: https:// www. pwc. com/gx/en/issues/data-and- analytics/publications/artificial-intelligence- study.html.

Rotman M. J., Data mining-a practical approach to database marketing (1995). IBM.

# Artificial Intelligent Tourism Behavioral Prediction Method

CHAPTER THREE

# Artificial intelligent tools predict travelling consumer behavior in airline and air agent travelling market

I believe that applying (AI) big data tool to predict vehicle buyer consumption choice behavior, it is similar to predict traveler consumption choice behavior. In this chapter, I shall indicate how to apply (AI) big data gathering tool to predict vehicle buyer consumption choice behavior. Then, I shall its what its similar points to be applied to predict traveler consumption choice behavior.

Nowadays, many vehicle manufacturers hope their vehicles can attract to vehicle buyers to choose to buy their vehicles. However, there are many different brands of vehicles to provide to them to choose, so the vehicle market competition is very serious.

How to judge their different kinds of vehicle price which is reasonable acceptance to attract vehicle buyers to choose to buy the brand of vehicle manufacturers‘ any kinds of vehicles, e.g. fast speed sport style vehicles, comfortable and slow speed common cars, for four passengers common small size or more than four passengers common large car size?

How to evaluate the vehicle prices issue is important factor to influence vehicle buyers' choices. Either if the brand of vehicle price is too high to compare other brands of similar vehicle price, it will influence many vehicle buyers choose to buy other brands‘ vehicles or if the brand of vehicle price is too low, it will influence vehicle buyers feel this brand's vehicle machine quality or safe driving level or manufacturing steel material or speed or not comfortable sitting etc. different factors is worse to compare to other vehicle brands' similar vehicle products.

Thus, if the brand of vehicle manufacturers can predict how to design vehicles which can attract many vehicle buyers to choose to buy whose any vehicle products. What are future vehicle buyers' favorable vehicle styles? Then, the vehicle manufacturer can concentrate on manufacturing the kind style of vehicle products to sell already. It will reduce its vehicle manufacturing investment risk.

How to apply (AI) tools to predict vehicle buyers' behavioral consumption model? Whether artificial intelligent tools can predict automotive buyers' behavioral consumption model and predict future vehicle design trend. In fact, automotive brands and dealerships are facing an increasingly competition when attempting to manually gathering the vast quantities of data required to create customer focused programs that increase retention, ultimately new sales and service automotive business.

Building a based on that client's intrinsic needs and interests to any kinds of automotive vehicles at any given time. This is especially true in the automotive industry where the time span between purchases is measured in years. Because vehicle buyers would not like often to change their old vehicle to another new one. So, their decisions to buying another new vehicle, the time is usually after one year, even longer time. Hence, it seems any vehicles won't be frequent consumption products to the owned at least one vehicle family consumers (vehicle buyers). It implies that why vehicle manufacturers ought need to spend time to predict future vehicle buyer design choice for whole year vehicle buyer number growth because they won't often change preferable vehicle design to change another new vehicle more easily.

Hence, how to predict vehicle consumers' taste or preferable which styles of vehicle choices issues is very important. If the vehicle manufacturers can not manufacture any attractive vehicles to sell easily in this year. Then, it will lose time, money in this year because it won't know when the owned least one vehicle users or non-owned any vehicle users who will decide to buy one new vehicle or change another new vehicle ensure. The different brand vehicle dealers will possible wait more than one year to attract them to buy their vehicles if their styles are not attractive to compare other brands of vehicle competitors.

However, artificial intelligence and machine learning can help any vehicle manufacturers to find solution to solve patterns in highly to solve patterns in highly complex data-sets that are beyond the capability of a human brain, and then building and automatically acting on the customer insights it generates.

Given the automotive customer need for individualized communications, this technology is positioned to become a critical component of any successful vehicle retailer's domestic or/and overseas vehicle markets. How can vehicle manufacturers and retailers use (AI) to enhance their vehicle marketing campaigns? How will (AI) affect their vehicle sale marketing strategy? What criteria would they use when selecting on (AI) solution?

Vehicle consumers today are able to quickly access different brands of vehicle information, research vehicle products and reviews, negotiate prices and compare one vehicle brand or retailer to another resulting of the brands of vehicle customers. At the same time, the rise of " big -data mining", wearable devices that track user's every move and preference and greater contextualization in advertising and social media has resulted in consumer expectations of individualized. Thus, it seems that (AI) tools can be used to gather " big-data" and then they can make human's mind to analyze how to design kinds of vehicles to satisfy vehicle buyers‘ needs.

As automotive vehicle marketers can apply (AI) tools to achieve messaging strategies to meet the needs of this new generation of informed vehicle consumers, using data from a variety of sources to move from a variety of sources to move from mass- messaging to more personalized messages aimed at particular vehicle buyer segments, e.g. fast speed sport vehicle buyer segment, slow speed comfortable small size or large size of buyer segment. However, when 90% of vehicle marketers believe having a single vehicle buyer view is important, only 6% have achieved it.

However, one of the main issues vehicle marketers are facing the lack of capacity to efficiently sift through and analyze the massive vehicle buyer amounts of data required to create vehicle buyer individualized vehicle customer experiences easily. This is especially difficult for automotive dealers, the long periods between purchase cycles, and the highly considered nature of the vehicle purchase means that each vehicle dealer needs to not only track a large number of potential vehicle customers for an extremely long period of time, but each of those vehicle customers will generate a huge amount of different kinds of vehicle behavioral consumption data as they research their next vehicle purchase. However, by choosing the right (AI) technological tools and programs , vehicle dealers can solve this big data gathering challenge into a major advantage.

For Forrester vehicle brand example, vehicle consumers have more power over the Forrester vehicle brand's reputation than ever before. Mayne, L. (2014) indicated that Forrester calls this new (AI) tools is the " age of the vehicle customer", a 20 year business cycle in which the most successful vehicle enterprises will reinvent themselves to systematically understand and serve increasingly powerful vehicle consumers. To win in this new age, Forrester declares companies must become vehicle customer obsessed and the only sustainable competitive advantage is knowledge and engagement with customers, such as (AI) gathering data knowledge.

Thus, the biggest challenge vehicle businesses currently face is not the collection of a large quantity of vehicle consumer data, but what to do with that data once they have it. Even at a large vehicle data research firm, the data sets are often too big for a single analyze, or even a team of analysts to sort through and draw conclusion from. However, enter artificial intelligence and machine learning , an efficient technology solution that can continuously find patterns in highly complex data sets that are way beyond the capacity of a human brain and then automatic drive action based on the customer insights is generated.

What is (AI) machine learning tool? Machine learning is a type of (AI) that learns from data and is not explicitly program. Think Amazon, face book. Machine learning serves up relevant content based on an individual vehicle purchase behavior and experiences. More simply, machine learning is a computer program that can learn relationships between data, subject those learnings to errors functions, and then learn from its errors. The program in effect, trains itself.

Lee, T. (2016) explained that "Thus, (AI) tools can learn deep a more advanced branch of machine learning inspired by how our brain's nervous function, has also been found to be especial effective in identifying patterns from data."

When this way sound is complicated from a vehicle dealer perspective, the implementation of a marketing program driven by artificial intelligence can take care of these tasks in an automatic vehicle fashion with little to no manual intervention required from the staff at time vehicle stores.

In practice at a vehicle dealership, the program will continue track vehicle customer behavior online, merging that data with any offline source ( like CRM or DMS data) and then analyze this aggregated vehicle buyer data set to predict what vehicle customer may be shopping for and what information they might like to relevance from different kinds style of vehicle design photos.

1.1 Why does travelling market seem to similar to vehicle market which can apply (AI) learning tool to predict travelling consumer behaviors?

Artificial intelligence refers to complex in vehicle market and travelling entertainment market which is very seem to be applied to predict consumer behaviors.

(AI) machine learning that posses the same characteristics of human intelligence and that have all our sense, all our reason and think just like human vehicle buyer who prefer vehicle purchase choice or travelling consumer who prefer travelling package or travelling destination and airline choice. Besides, machine learning is the practice of using algorithms to collect and examine data, learn from it, and then make a determination or prediction about something in the world.

So, it can be attempted to gather data concerns that travelling consumer past travelling destination choice and air ticket price choice and different travelling package, e.g. high, middle, or low class hotel and foods supply and entertainment places choice in their past travelling journeys.

The machine is " trained" using large amounts of data and algorithms that give it the ability to learn how to automatically perform a task with increasing accuracy. Otherwise, deep learning is primarily based on artificial neural networks inspired by our understanding of the biology of human's brains.

Thus, (AI) big data can gather all these past traveler consumption behavioral choice data to make reference to analyze whether how many travelers will choose to go to the specific travelling destination in any time by the past traveler number record to different travelling destinations, then it can gather the past air ticket sale price to different destinations and past travelling package design to different destinations in order to analyze whether it is the cheap airline ticket price factor or attractive travelling package factor or attractive travelling entertainment etc. in order to predict which factor is the most potential influential factor to they choose to go to the destination to travel in different time within one year. Then, traveler agent or airline can collect these big data to judge how to design their package to attract travelers to go to anywhere to travel or what the main factor influence most of them to choose to visit the destination to travel.

For example, travel agents or airlines can apply "Deep learning" breaks down tasks in ways that enables machines to assist them to predict when travelling consumer choice will be changed and why their travelling choice will change and how their travelling choice will change with increasingly complex tasks.

So, such as why (AI) technology can be applied to predict how travelling consumer behavior changes to bring to judge whether anywhere will be many travelling consumers who will prefer to choose travelling hot destinations next year or next month.

Then, travel agents and airlines can gather overall past travelling consumer data to analyze and conclude the more accurate prediction of different travelling destinations to the number of traveler. Then, they can choose how much air ticket price is more reasonable to charge to the travelling destination or how to design the travelling package which can bring more attractive to the prediction number of different travelling destination travelers in order to achieve to raise the different travelling destination number next year.

Thus, (AI) big data machine learning can help airlines or travel agents to solve how to design any attractive travelling package challenge. A travelling package is both one of the most important and carefully considered travelling entertainment consumption the majority of travelling people will ever make in their lifetime at least one travelling time.

It is also a prediction how travelling package will be designed that tends to be fundamentally tied to a travelling person's travelling destination choice identify and travelling package view of themselves. As the same time, travelling consumers' travelling choice changing lifestyles result in changing travelling destination needs, e.g. the country's young travelers can choose to change non-extreme exciting travelling entertainment package from past extreme exciting travelling entertainment package. Due to personal feeling factor in general. However, I believe that (AI) big data can also be attempted to predict when the country's young travelers will choose to change non-extreme exciting travelling behavior.

It is similar to automotive dealers need to remember that vehicle customers and prospects are individual human beings with risk, complex and ever-changing lives factors, these factors will influence every vehicle consumer why who feels has vehicle purchase need, and how who choose to buy the first vehicle if who decided to buy the first vehicle.

It seems that travelling agents or airlines need to remember that travelling consumers and different features or designs are very traveler beings with risk, complex and ever-travelling package attitude personal changing factors in different travel season, these factor will influence every individual traveler why who feels has travel entertainment need, and how who choose to buy different feature or design travelling package if who decide to travel. The (AI) big data technological travelling customer behavioral prediction tool seems to be the best travelling behavioral prediction tool in the world are those that know every one of different country's traveler need. Their likes and dislikes which style of travelling package, preferences and travel destination changing tastes to travelling destination choices.

The capacity of the human brain, however, limits us from achieving these different type of travel package sales. In this competitive travelling destination choice entertainment environment, (AI) big data machine learning enables platforms to assist the air ticket and travel package sales team by tracking the travelling consumer behaviors of each travelling customer, learning and memorizing their preferences and predicting their future travelling destination choice and travelling package design needs.

Finally, I recommend that for a travel agent or airline travelling marketing platform to make their travelling customer engagement efficient and fully-functional, I should be able to: applying (AI) tools to track every travelling customer behavior across the web, connecting to a society of data sources, CRM, DMS, third-party, web travelling brands, social traveler email, click etc., aggregating and accurately cross-reference data from a variety of sources, leveraging this data to drive insights on a mass scale, as well as on an individualized basis, driving actions and automatically direct travelling customer engagement via multiple channels based on where each customer is in their travelling individual lifecycle.

1.2 Why is (AI) big data gathering tool better than psychological and survey methods to predict traveler individual travel choice behavior?

Prediction travel behavioral consumption from psychology and survey methods.

How to predict travel consumption? It is one question to any travel agents concern to use what methods which can predict how many numbers of travelers where who will choose to go to travel more accurately. I think that who can consider how to predict travel behavioral consumption from psychology and survey travel choice prediction method, but it is better to apply (AI) big data gathering method to predict travel consumer's destination choice more accurate. The reason is as below:

The first reason is that traveller individual travel psychological desire is difficult to predict accurate more than (AI) big data gathering method, it is due that the data is past traveler's destination choice and travel package and ticket price actual data from (AI) big data gathering method. Otherwise, survey investigation is only traveler psychological thinking method. It lacks enough past actual traveler data gathering.

The second reason is that on the weakness of traveler individual psychological thinking view of survey investigation. It has evidence to support the relationship between self-identify threat and resistance to change travel behavior to any travelers, controlling for whose past travelling behavior, resistance to change if a psychological phenomenon of long standing interest in many applied branches of psychology.

Past travelling behavior has been acknowledged as a predictor of future action. Such as travelling behavior that is experienced as successful is likely to be repeated and may lead to habitual patterns. Some psychologists differentiate habit between two concepts, such as goal oriented and automatic oriented both. Although repeated past travelling behavior is addition goal oriented and automatic oriented. Further non-deliberative nature of habit may make appeals to judge and to predict future individual traveler's behavior accurately.

However, repeated one traveler will choose the destination to repeat to travel without a necessary constraint of goal orientation and automatic oriented both. So, it seems that psychological factor can influence any individual traveler why and how who choose to decide to repeat to choose the destination to travel.

So, survey investigation is only the traveler's thinking to answer the travel firm. It is not sure that the traveler's past travel experience is real answer. Otherwise, (AI) big data gathering method is computer gathering method which gather past traveler consumption actual data to analyze and conclude future traveler possible repeated travel destination choice and travel package choice more accurate.

The third reason is that on the strength of (AI) big data gathering method computer statistic view to predict future traveller consumer's destination and travel package choice. It is structural equation modeling is an extremely flexible linear-in-parameters multivariate statistical modeling technique. It has been used in modeling travel behavior and values since about 1980 year. It is a software method to handle a large number of variables, as well as unobserved variables specified as linear combinations ( weighted averages) of the observed variable.

1.3 Can (AI) big data gather data to predict when climate will change to influence poor travelling behaviours?

(AI) big data tool can predict the flexibility of human travelling behavioral change is at least the result of one such mechanism, our ability to travel mentally in time and entertain potential future. Understanding of the impacts is holidays, particularly those involving travel.

Using focus groups research to explores tourists' awareness of the impacts of travel own climate change, examines the extent to which climate change features in holiday travel decisions and identifies some of the barriers to the adoption of less carbon intensive tourism practices.

The findings suggest many tourists don't consider climate change when planning their holidays. The failure of tourists to engage with the climate change to impact of holidays, combined with significant barriers to behavioral change, presents a considerable challenge in the tourism industry. In the future, computer (AI) big data tool can attempt to predict when the country's climate change to influence travelers to choose to go to the country to travel, e.g. next month or next half year or next year hot travelling destinations.

Tourism is a highly energy intensive industry and has only recently attracted attention as an important contributions to climate change through greenhouse gas emissions. It has been estimated that tourism contributes 5% of global carbon dioxide emissions. There have been a number of potential changes proposed for reducing the impact of air travel on climate change. These include technological changes, market based changes and behavioral changes.

However, the role that climate change plays in the holiday and travel decisions of global tourists. How the global tourists of the impacts travel has on climate change to establish the extent to which climate change, considerations features in holiday travel decision making processes and to investigate the major barriers to global tourists adopting less carbon intensive travel practices.

It will bring this question: Will tourists aware the impacts that their holidays and travel have on climate changes to influence their travelling decision?

When, it comes to understand individual traveler's behavioral change, wide range of conceptual theories have been developed, utilizing various social, psychological, subjective and objective variables in order to model travel consumption behavior. These theories of travel behavioral change operate at a number of different levels, including the individual level, the interpersonal level and community level. Whether pro-environmental behavior can be used to predict travel consumption behavior in a climate change. However, the question of what determines pro-environmental behavior in such a complex one that it can not be visualized through one single framework or diagram.

Despite the potentially high risk scenario for the tourism industry and the global environment, the tourism and climate change ought have close relationship.

However, (AI) big data tool can be applied to find what factors to influence the time of travelers' travelling choices. What are the important factors and variables which can limit tourism? e.g. money, time, family problem, extreme hot or cold weather change, air ticket price, journey attraction etc. variable factors.

Mention of holidays and travel were deliberately avoided in the recruitment process, so as not to create a connection factor to influence traveler's individual mind. However, the dismissal of alternative transportation modes can be conceived as either a structural barrier, in the sense that flying is perhaps the only realistic option to reach long-haul holiday destination, or a perceived behavioral control barriers in that an individual perceives flying as the only option open to whom.

The transportation tool factor will be depend to extent on the distance to the destination. This can also be interpreted in a social perspective as an intention with the resources available where much international tourism is structured around flying. To increase the availability of different transportation modes, tourists could choose holiday destination closer to home.

Finally, also how to predict future travel behavioral consumption. I feel that travel agents need to predict whether any country's random daily variation of weather factor is also important to influence travel behavior. e.g. in weather, temperature, rainfall and snowfall with traffic accidents factors will have relationship to cause travel demand.

Some scientists estimate suggest that when warmed temperatures and reduced snowfall are associated with a moderate decline in non-fatal accidents, they are also associated with a significant increase in fatal accidents. Thus increase in fatalities and temperature. Half of the estimated effect of temperature on fatalities is due to changes in the exposure to pedestrians, bicyclists and motorcyclists as temperature increase.

So, if any countries have rainfall, snowfall and low temperature to cause traffic accidents, whether this accident occurrence will influence the travelers who liking climb snow hills, riding bicycle, running sports who will avoid to travel to these countries‘ bad weather after occurs. So, why I feel that this natural climate factor will also be one serious factor to influence travel behavioral consumption. However, (AI) big data tool can predict more accurate than survey method when climate change to influence the country's climate to be poor, then it can predict when which countries are not popular acceptable to global country consumers' travel choice next month.

CHAPTER FIVE

# (AI) predicts traveller behaviors with better-informed decisions

I shall explain how (AI) big data gathering technology can provide travelling businesses with better-informed decisions to drive top-line growth, deliver meaningful experience for travelling customers and smooth their path along the travelling consumer journey. The widely understood definition of (AI) involves the ability of machines or computers to learn human thinking, reasoning and decision-making abilities.

So, such as (AI) learning machine system can attempt to learn travelling consumer's travel destination or travel package thinking, judgement of their reasons why they choose to go to the destination to travel or why they choose to buy the travel package and learn how and why they make their past travelling decisions from their past travel big data gathering.

A Narrative science study in 2015 year identified that (AI) was being used primarily in voice recognition, machine learning virtual assistants and decision support. This study also highlighted the many branches of (AI) and that techniques and their definition are used interchangeably. It is possible that (AI) can be used to gather big data , then to analyze to help travel businesses to predict travelling consumer travel destination and travel package choice behaviors. For example, one of the most common techniques is traveler machine learning, where algorithms are used to perform tasks by learning from the airline or travel agent whose past all travelers‘ travelling destination choice and travel package choice historical data.

However, during 2017 year, search engines will begin to find what additional factors can influence past traveler personal travelling destination and travelling package travelling behavioral data into prediction of future travelling customer behavioral results, such as the online traveler (user's) history of travelling data searches, such as anywhere are the most popular travelling locations or travelling destinations and previously captures conservations.

Artificial intelligence will use this past travelling destinations and travelling package information to power predictive search results, e.g. predictive future travelling consumer's choice behavioral processing for where will be their preferable travelling destination choice and how to design travelling package to satisfy future travelling clients' needs.

Predictive search will improve the quality of online travelling search results, and provide new insights into travelling consumers' travelling destination and package behavior and the moments which matter to them. Search will give recommendation into tailored how travelling consumer individual travelling destination choice in travelling decision making process. Several of the largest online platforms already use (AI) travelling machine learning to improve predictive travelling consumer behavioral search results.

For example, Google's rank brain technology adds research by understanding the context in which the travelling consumer has entered it. Over time, rank brain will learn further from user behaviors Amazon's DSSTNE ( pronouned destiny) learns from shoppers' purchasing habits and consumption behavior to offer better product recommend actions, which Amazon can offer before a consumer has entered anything into the search bar.

Such as (AI) big data can gather past online travelers' e-ticket purchase transactions to conclude that online traveler's travelling choice habits and online traveler consumption behavior to offer better travelling destinations and travelling package opinions to travel agents or airlines. However, this technology is not independent of human input. For example, Google engineers will periodically retain the rank brain system to improve the models it uses.

For another example, in 2016 year , Apple computer revamped its travelling scene photos app to allow travelling consumers to search for specific travelling destinations in the travelling scene phots, they want to find anywhere travelling destination photos, not just dates and locations. Each travelling photo that an intelligent phone or intelligent pad user takes goes through 11 billion computations, so that travelling scene photos can understand exactly where is the travelling destination photography to let online travelling consumer to feel anywhere they plan to go to the location to travel. So, (AI) learning machine can make online travelling photos more attractive to influence potential travelers choose to the destination to travel after they see the travelling destination scene photos from internet.

It seems that in future, (AI) machine learning will allow online travelling search to evolve even further. Search engineers will deliver refined recommendations to airlines' online traveler e-ticket search users and use less human input to predict travelling consumers' needs from internet channel. For IBM computer example, it indicated 90% of the data that exists today has been created in the last two years.

This huge explosion of past traveler's e-ticket consumption data gives the opportunity to quickly spot and react to the latest trends, fashion and fads among its travelling clients and potential clients. This will allow airline or travel agent companies to better engage with younger travelling consumers, who gain influence access to the latest travelling destination and package trends.

They associate with to help define who they are as individuals. Thus, travelling company brands have to identify and make use of them before travelling consumers move on, but the vast quantity of past e-ticket purchase data available makes from internet channel. This a resource-intensive task. For next example, Lesara, a based online clothes store, uses this machine learning to inform its product decision often gathering information from internal and external sources.

When its trends -spotting shoes. Lesara has a range of over 20 styles and sells hundreds of pairs a day. It focus on giving consumers, the very latest trends allow Lesara to develop on average of 50,000 new items each year. It compared to 11,000 old items each year. Thus, travelling agents or airlines can attempt to apply (AI) big data gathering method to gather all past e-ticket purchase data, concerns where they prefer to choose to go to the destinations to travel and what travelling packages are the most attractive to the travelers to choose to buy. It aims to help them to predict where future travelers will prefer to choose to go to travel or what travelling package they will prefer to choose to buy next year.

For another (AI) big data prediction example, Lesara is one online clothes store, uses machine learning decisions after gathering information from internal and external sources. One of its most popular products, shoes with LED started life when its trend spotting software flagged up a blogger wearing similar shoes. Now Lesara has a range of over 20 styles and sells hundreds of pairs a day. Its focus on giving consumers the very latest trends allows Lesara to develop an average of 50,000 new items each year, compared to 11,000 for its competitor Lara.

It seems (AI) big data gathering machine learning can help Lesara business to predict what kinds of shoes design or style that shoe consumers will prefer choose to buy in future shoe market trend. Thus, Lesara can predict shoe consumers' taste successfully and it can manufacture many attractive style of shoes.

(AI) machine learning can gather global past shoe consumer's shoe shopping experiences, then analyzes to make conclusion to give lesara recommendation successfully. This will make the experience more enjoyable for shoe consumers and allow Lesara to advert whose different new style or design of shoes to deliver them move relevant messages by understanding the context of the experience.

So, online travel agents or online airline can also attempt to apply (AI) big data gathering method to predict where travelers will prefer to go to travel and how they ought design travelling packages to attract them to choose to buy next year. Hence, (AI) big data gathering technology can conclude how to design traveler agents' travelling package products to be the most attractive to excite many travelers choose to buy their travelling package, due to it has more accurate to predict travelling consumer destination and travelling package choice behaviors to compare human themselves prediction judgement effort, e.g. travelling survey or marketing research, or telephone enquire. It seems that (AI) machine judgement effort is more accurate to compare to human judgment effort in travelling industry.

## 2.1 Future travel consumption behavior

Can (AI) big data gathering tool predict traveler individual habitual behavior , e.g. renting travel transportation tools ?

Can (AI) big data gathering tool can predict past traveler destination and travelling package choice habit and it can be intended to predict of future traveler behavior to people are creatures of habits judgement of future anywhere travelling destination choice next year or next month or next half year destination prediction ?

Many of human's everyday goal-directed behaviors are performed in a habitual fashion, the transportation made and route one takes to work, one's choice of breakfast. Habits are formed when using the some behavior frequently and a similar consistency in a similar context for the some purpose whether the individual past travel consumption model will be caused a habit to whom. e.g. choosing whom travel agent to buy air ticket or traveling package; choosing the same or similar countries' destinations to go to travel ; choosing the business class or normal (general) class of quality airlines to catch planes.

Does habitual rent traveling car tools use not lead to more resistance to change of travel mode? It has been argued that past behavior is the best predictor of future behavior to travel consumption. If individual traveler's past consumption behavior was always reasoned, then frequency of prior travel consumption behavior should only have an indirect link to the individual traveler's behavior. It seems that renting travel car tools to use is a habit example. So, a strong rent traveling car tools useful habit makes traveling mode choice. People with a strong renting of traveling car tools of habit should have low motivation to attend to gather any information about public transportation in their choice of travelling country for individual or family or friends members during their traveling journeys.

Even when persuasive communication changes the traveler whose attitudes and intention, in the case of individual traveler or family travelers with a strong renting travel car tools habit. It is difficult to change whose travel behaviors to choose to catch public transportation in whose any trips in any countries. However, understanding of travel behavior and the reasons for choosing one mode of transportation over another. The arguments for rent traveling car tools to use, including convenience, speed, comfort and individual freedom and well known.

Increasingly, psychological factors include such as, perceptions, identity, social norms and habit are being used to understand travel mode choice. Whether how many travel consumers will choose to rent traveling car tools during their trips in any countries. It is difficult to estimate the numbers. As the average level of renting travel car tools of dependence or attitudes to certain travel package policies from travel agents. Instead different people must be treated in different ways because who are motivated in different ways and who are motivated by different travel package policies ways from travel agents.

In conclusion, the factors influence whose traveler's individual traveler destination choice behavior The factors include either who chooses to rent traveling car tools or who chooses to catch public transportation when who individual goes to travel in alone trip or family trip. It include influence mode choice factors, such as social psychology factor and marketing on segmentation factor both to influence whose transportation choice of behavior in whose trip. So, (AI) big data can be attempted to gather past traveler transportation tool choice, rent travelling car tools choice or catching public transportation tools choice to predict where destination can provide what kind of transportation tool to attract many travelers to choose to go to the place to travel.

How (AI) big data determine future travel behavior from past travel experience and perceptions of risk and safety for the benefits to travel consumers?

How (AI) big data determine future travel behavior from past travel experience and perceptions of risk and safety for the benefits to travel consumers? Why does individual traveler avoid certain destination(s) is(are) as relevant to tourist decision making as why who chooses to travel to others?

Perceptions of risk and safety and travel experience are likely to influence travel decisions. If travel agents had efforts to predict future travel behavior to guess whether travelers will feel where is(are) risk and unsafe to cause who does not choose to go to the country to travel. Then, the travel agents will avoid to choose to spend much time to design the different traveling package to attract their potential travel consumers to choose to travel. The reason is because in the case of individual traveler's tourism experience, the traveler whose past disappointment travel experience ( psychological risk) will be a serious threat to the traveler's health or life ( health, physical or terrorism risk). The past safety or unhealthy risk to the country(countries) will influence the traveler decides to choose not to go to the countries(country) to travel again in the future.

2.2 What is push and pull factors to influence any traveler who chooses where is whose preferable travelling destination ?

How to apply (AI) big data to predict individual traveler's behavioral intention of choosing a travel destination?

Understanding why people travel and what factors influence their behavioral intention of choosing a travel destination is beneficial to tourism planning and marketing. In general, an individual's choice of a travel destination into two forces.

The first force is the push factor that pushes an individual away from home and attempt to develop a general desire to go somewhere, without specifying where that may be.

The other force is the pull factor that pull an individual toward in destination, due to a region-specific or perceived attractiveness of a destination. The respective push and pull factors illustrate that people travel because who are pushed by whose internal motives and pulled by external forced of a destination. However, the decision making process leading to the choice of a travel destination is a very complex process.

For example, a Taiwanese traveler who might either choose new travel destination of Hong Kong or another old travel Asia destinations again or who also might choose any one of Western country, as a new travel destination. The travel agents can predict where who will have intention to choose to travel from whose past behavior and attitude, subjective and perceived behavioral control model. When (AI) big data gather past every country traveler number who chose to go to which countries to travel in order to judge where destinations will be the country travelers' travelling choice destinations in the future.

The factors influence where is the traveler choice, include personal safety, scenic beauty, cultural interest, climate changing, transportation tools, friendliness of local people, price of trip, trip package service in hotels and restaurants, quality and variety of food and shopping facilities and services etc. needs. So, whose factors will influence where is the individual travel's choice. It seems every traveler whose choice of travel process, will include past behavior. e.g. travelling experience, travelling habit, then to choose the best seasoned travelling action to satisfy whose travel needs. This process is the individual traveler's psychological choice process, who must need time to gather information to compare concerning of different travel packages, destination scene, climate change, transportation tools available to the destination, air ticket price etc. these factors, then to judge where is the best right destination to travel in the right time.

Hence, (AI) big data can gather past different countries' climate changing data, transportation tool changing data, destination scene environment changing etc. different data to give opinions to travelling businesses whether any country's these above factors will influence about how many traveler number will be increase or decrease in the future.

2.3 Why can expectation, motivation and attitude factor influence travelling behavior?

Social psychology is concerned with gaining insight into the psychological of socially relevant behaviors and the processes. For instance, on a global level bad influence to global warming, it influences some countries extreme cold or hot bad climate changing occurrence, then it ought influence some travelers' behavioral decision to change their mind to choose some countries to go to travel at the moment which do not occur extreme hot or cold climate ( temperature). e.g. above than 40 degree in summer or below than 0 degree in winter. Due to the extreme climate changing environment in the countries, it will cause them to feel uncomfortable to play during their trips. So, the global warming causes to climate changing factor will influence the numbers of travel consumption to be reduced possibly. This is global climate changing environment factor influences to bad or uncomfortable social psychological feeling to global travelers' mind of traveling decision. What is individual traveler expectation, motivation and attitude? Tourism sector includes inbound (domestic) tourism and outbound (overseas) tourism both incomes to any countries. According to recent article, a tourist behavior model has been developed, called the expectation, motivation and attitude ( EMA) model ( Hsu et al., 2010).

This model focuses on the pre-visit stage of tourists by modeling the behavioral process by incorporating expectation, motivation and attitude. Travel motivation is considered as an essential component of the behavioral process, which has been increasing attention from the travel; industry. The economic approach defines "tourism" is an identifiable nationally important industry. It includes the component activities of transportation, accommodation, recreation, food and related service. So, tourism behavioral consumption is concerned the individual tourist's usual habituate of the industry which responds to whose needs, and of the impacts that both the tourist and the tourism industry have on the socio-cultural, economic and physical environment.

However, travel motivation means how to understand and predict factors that influence travel decision making. According to Backman and others (1995, p.15), motivation is conceptually viewed as " a state of need, a condition that services as a driving force to display different kind of behavior toward certain types of activities, developing preferences, arriving at some expected satisfactory outcome." So, motivation and expectancy which has close relationship to any tourist before who decided to do any tourism of behavior.

Some economists confirmed motivation and expectancy which has relations, such as expectation of visiting an outbound destination has a direct effect on motivation to visit the destination; motivation has a direct effect on attitude toward visiting the destination; expectation of visiting the outbound destination has a direct affection on attitude toward visiting the destination and motivation has a mediating effect on the relationship in between expectation and attitude.

Hence, (AI) big data can gather all the country's climate environment change, transportation tool change, entertainment scene change, hotel price and restaurant price change etc. data to give opinions whether the country will attract how many traveler to choose to go to travel in the year.

CHAPTER SIX

# (AI) deep learning techniques to forecast travelling environment behavioral consumption

environment behavioral consumption
Prediction how many travelers will choose to go to the country to travel. It is similar to apply deep-learning technology to predict how to raise the agricultural farming productivity in the agricultural export country.
The (AI) deep-learning technology leads to performance enhancement and generalization of artificial intelligent technology. It influences the global leader in the field of information technology has declared its intention to utilize the deep-learning technology to solve environmental problems, such as climate change.
So, it will help agriculture farming businesses can raise any plant food: vegetable, fruit, rice which grow up very easily if farmers can apply (AI) deep-learning technology to solve environment problems to influence their plant food grow. If the whole year seasonal change is very good and it is suitable for any plant food to grow in farming land easily, e.g. rain is enough and soil is enough for any plant food to grow in the farm lands. Then, fruit, rice, vegetable etc. agriculture businesses will have much beneficial attribution to global farmers.
The question is how to use deep-learning technologies in the environmental field to predict the status of pro-environmental consumption. We predicted the pro-environmental consumption index based on Google search query data, using a recurrent neural network ( RNN model). To certify the

accuracy of the index, we compared the prediction accuracy of the RNN model with that of the ordinary least square and artificial necessary network models.

For example, the RNN model predicts the pro-environmental consumption index better than any other model. we expect the RNN model to perform still better in a big data environment because the deep-learning technologies would be increasingly as the volume of data grows. So, deep-learning technologies could be useful in environmental forecasting to prevent damage caused by climate change to influence any rice, vegetable, tomato, potato, fruit etc. different plant food grow in any countries' farming land easily.

For South Korea example, over 800 government agencies spent 2.2 trillion Korea won on eco-products in 2014 year. However, green products are rarely purchased outside these agencies. This phenomenon occurs because there is a gap between consumer attitudes and behavior , that is environmental attitude is a major factor in decision making vis-a-vis the consumption of " green" food and services ( Jorea Ministry of Environment, 2015).

Therefore, it is necessary to understand those consumer attitude, that will lead to sustainability-conductive behavior and consumption. (AI) Deep learning system can be applied to attempt understand those traveler attitude to environment protection to fly to which country. For example, (AI) deep learning system can attempt to gather data concerns how many Hong Kong people concern air pollution challenge to influence their health, then it can attempt to predict how many Hong Kong travelers do not choose to go China travel, due to the air pollution challenge to influence their health.

3.1 Environmental travel consumption prediction

Recently, many researchers have studied pro-environmental consumption and household indexes as well as suicide rate predictions using messages posted by internet users on Google trend, Tweets etc. channel.

Whether can environmental consumption be predicted by (AI) deep-learning technological internet channel to influence how many travelers choose to go to the country to travel?

How can impact the pro-environmental consumption attitudes of green policies to influence how many travelers choose to go to the country to travel?

For example, Korea scientists estimated pro-environmental attitudes using search query data provided by Google trend and confirmed through regression analysis, that pro-environmental attitude has a positive correlation with the pro-environmental attitude index. They also explained that environment-friendly attitude of residents plan an important role in policy making. In the past, most household consumption indexed were calculated through surveys, but (AI) deep-learning technological tool " big data" have recently gained research attention ( Lee et al. 2016). So, (AI) deep learning technology can attempt to gather whether how many Korea residents who concern environment pollution to influence their eating green food attitude then to judge whether how many Korea residents hope to leave their country to travel anywhere either high risk environment pollution countries to travel or low risk environment pollution countries to travel in the future.

It seems that (AI) deep-learning technology can help agricultural export countries' farmers , e.g. US, UK, Canada, New Zealand, Australia, Japan, China, India etc. they can predict environmental behavioral consumption to any rice, tomato, potato , fruit, vegetable etc. plant food consumers. The beneficial advantages to them include as below:

(a) Assuming they know their countries' weather, when it has less rain to cause drought or when it has more rain in any seasonal time in the year. They can choose not to grow any kinds of above these plant food to avoid loss.

(b) They can make any kinds of above these plant food price raising after their prediction of these bad seasonal time to cause their plant food shortage supply challenge. Because these plant food consumers' demand number is more, but the supply of these above plant food supply number is less. However, due to they had predicted when the bad seasonal time can not allow them to grow these above plant food before. So, they have enough time to grow many these above plant food number in predictive good seasonal time to prepare to supply to their plant food import countries' plant food consumers to eat. Thus, these predictive environmental consumption plant food export countries can raise their plant food price to sell to them. When, the other non-pre-predictive environmental consumption plant food export countries can not supply any one of those plant food to them to eat, due to the bad climate to cause them can't grow any one of these plant food to export to sell.

Thus, (AI) deep-learning technology can be applied to predict how to raise the plant food supply number in order to raise price to the import plant food countries consumers to eat, due to they feel difficult to buy these plant food to eat in the bad climate seasonal time in whole year.

(c) (AI) deep-learning technology can help climate scientists to find what reasons cause their countries; rain sudden increases or cause their countries' rain sudden decreases. After its gathering data analysis, it can assist climate scientists to find solution methods to attempt to control the rain level can be right falling down level to let agricultural export farmers who can grow their plant food to sell to agricultural import countries in whole year.

(d) The agricultural export countries' farmers can apply (AI) deep-learning technology to help them to choose whether growing which kinds of plant food in that whether climate time to earn more plant food consumption number more easily.

Due to the agricultural countries climate will often change, for example, tomato, potato, rice, fruit etc. plant food can be adapt to grow in more rain time, but vegetable can not be adapt to grow in more rain time. If farmers can apply this technology to predict when it will have move rain or when it will have less rain to fall down in their countries. Then, they can choose to grow which kinds of plant food number more, in the suitable seasonal climate time in order to raise plant food growing number productivities to supply to sell to satisfy any agricultural food import countries' demand effectively.

(e) (AI) deep-learning technology can help agricultural import countries to solve agricultural food shortage challenge in long term. When this technology can be popular to base applied by the agricultural plant food export countries. It will solve global agricultural food shortage challenge. For example, when one agricultural export countries' farmers can popular accept to apply this technology to predict when to grow which kinds of plant food more to rise number productivities to sell. e.g. vegetable, fruit, rice Besides another agricultural export countries' farmers can also accept to apply this technology to predict when to grow plant food, e.g. potato, tomato to raise number productivities to sell. Then, they can concentrate on growing the specific kinds of plant food in order to raise the specific plant food number productivities in every seasonal change time every month. Then, global agricultural plant food supply must be raised, due to these predictive environmental change farmers can know who ought grow which kinds of plant food to sell to raise number productivities.

Consequently, (AI) deep learning can gather where countries will have high risk environment pollution to influence health food supply. Then, it can give opinions to travelling businesses when these high risk environment pollution countries will encounter the traveler number to be decreased, due to the environment pollution serious challenge will occur.

3.2 What methods can predict future travel behavioral consumption ?

How to use qualitative of travel behavioral method to predict future travel consumption from (AI) big data ?

I also suggest to use qualitative of travel behavioral method to predict future travel consumption. Methods such as focus groups interviews and participant observer techniques can be used with quantitative approaches on their own to fill the gaps left by quantitative techniques. These insights have contributed to the development of increasingly sophisticated models to forecast travel behavior and predict changes in behavior in response to change in the transportation system. I shall indicate the weaknesses of human travelling investigation methods as below:

First, survey methods restrict not only the question frame but the answer frame as well, anticipating the important issues and questions and the responses. However, these surveys methods are not well suited to exploratory areas of research where issues remain unidentified and the researched seek to answer the question "why?".

Second, data collection methods using traditional travel diaries or telephone recruitment can under represent certain segments of the population, particularly the older persons with little education, minorities and the poor.

Before the survey, focus group for example can be used to identify what socio-demographic variables to include in the survey, how best to structure the diary, even what incentives will be most effective in increasing the response rate.

After the survey, focus, focus groups can be used to build explanations for the survey results to identify the "why" of the results as well as the implications. One Asia Pacific survey research result was made by tourism market investigation before. It indicated the travel in Asia Pacific market in the past, had often been undertaken in large groups through leisure package sold in bulk, or in large organized business groups, future travelers will be in smaller groups or alone, and for a much wider range of reasons.

Significant new traveler segments, such as female business traveler. The small business traveler and the senior traveler, all of which have different aspirations and requirements from the travel experience.

Moreover, Asia tourism market will start to exist behaviors in the adoption of newer technologies, a giving the traveler new ways to manage the travel experience, creating new behaviors. This with provide new opportunities for travel providers. The use of mobile devices, smartphones, tablets etc. and social media are the obvious findings to become an integral part of the travel experience. Thus, quality method can attempt to predict Asia Pacific tourism market development in the future. It is such as (AI) big data gathering tool can give traveler quality opinions to any travelling businesses to make the more accurate where will be the popular travel destination choice next month or next half year or next year.

However, improving the predictive power of travel behavior models and to increase understanding travel behavior which lies in the use of panel data( repeated measures from the same individuals). Whereas, cross-sectional data only reveal inter-individual differences at one moment in time, panel data can reveal intra-individual changes over time. In effect, panel data are generally better suited to understand and predict ( changes in ) travel behavior. However, a substantial proportion was also observed to transition between very different activity/travel patterns over time, indicating that from one year to the next, many people renegotiated their activity/travel patterns.

3.3 How to apply advanced traveler information systems (ATIS) to predict future travelling behavior?

Nowadays, information can impact on traveler behavior and network performance. For example, when steadily growing levels of vehicle ownership and vehicle miles traveled information has been identified as a potential strategy towards man aging travel demand, optimizing transportation networks and better utilizing available capacity. Toward, this goal to predict further tourist behavioral consumption. Many countries, government tourism development institutes has applied advanced traveler information systems (ATIS) which travel behavior models and high-fidelity network performance models made increasingly feasible through the rapid advances in computer power. Crucial components of this problem domain are the modeling of individual tourist drivers' response to travel information and the development accurate guidance of relevance to real would trip makers. So, this advanced traveler information systems (ATIS) can assist the tourist who like to rent travelling car tools to travel in any countries own free traveler information systems service conveniently. Also, this travel information system can be intended to assist travelers to make better travel choices. e.g. this system can improve the decision making of individual traveler rather than improvements of network performance overall. So, we need to understand how tourists make their travel plans. Also, understanding decision process that lead to booking of the trip is equally important, as it allows of a potential behavior.

3.4 How can online tourism sale channel influence traveling consumption of behavior?

Nowadays, internet is popular, it seems that booking air ticket behavior of using internet is predicted to influence overall tourism air tickets payment method. Tourism industry has grown in the previous several decades. Despite its global impact, questions related to better understanding of tourists and whose habits. Using online travel air ticket booking benefits include booking electronic air tickets can be made from entering any electronic travel agents websites in the short time and electronic travel ticket payers do not need leave home, who can pay visa card to pre booking any electronic travel ticket from online channel conveniently.

3.5 How can analyze activity based travel demand ?

Nowadays, human are concerning the traffic congestion and air quality deterioration, the supply oriented focus of transportation planning has expanded to include how to manage travel demand within the available transportation supply. Consequently, there has been an increasing interest in travel demand management strategies, such as congestion pricing that attempts to change aggregate travel demand. The prediction aggregate level, long term travel demand to understanding disaggregate level ( i.e. individual levels ) behavioral responses to short term demand policies, such as ride sharing incentives, congestion pricing and employer based demand management schemes, alternate work schedules, telecommuting limitation of travel agent traditionally work nature shall influence oriented trip based travel modelling passenger travel demand indirectly.

Finally, online travel purchase will be popular to influence the number of travel behavioral consumption nowadays. Any travel package products can be sold from websites to attract travelers to choose to pre-book air ticket for any trips conveniently. In the past ten years, the internet has become the predominant carrier of all types of information and transactions. Regarding travel decisions, internet has also become an important sales channels for the travel industry, because it is associated with comparably lower distribution and sales costs, but also because it adapts to high supply and demand dynamics in this industry. Consequently, the travel and tourism industry tries to increase the internet sale specific share of sales volumes. So, internet sale channel has changed travel consumption behavioral pattern and characteristics and travel experience. For example, Switzerland has one of the highest population-to-computer ratio in Europe. It is also one of the most highly internet penetrated countries in terms of use of the WWW on a day-to-day basis, with more than 75 percent of the population older than 14 years using the WWW daily ( ICT, 2005).

The reason of booking online tourism may include: convenience, fast transaction, finding traveling package choice easily, more airline seats available. So, online booking tourism will influence the traditional tourism agents visiting of sales and air tickets and travelling package numbers to be decreased. Finally, the online booking tourism market shares will be expanded to more than traditional tourism agents visits sale market in the future one day. So, the travel agents who still use the traditional tourism visiting sale channel which ought raise whose features to compare to differ to online tourism sale channel if these traditional tourism agents want to keep competitive ability in tourism industry for long term.

3.6 What is actively based patterns of urban population of travel behavioral prediction method?

Actively based patterns of urban population. It is a method of motivational framework means in which societal constraints and inherent individual motivations interact to shape activity participation patterns. It can be used to predict one city or urban the numbers of travel demand in the year. It has two elements: First, capability constraints refer to constraints are imposed by biological needs, such as eating and sleeping and/or resources, such as income, availability of cars etc. to undertake the urban or city's family activities in the year. Second, coupling constraints define where, when and the duration of planning activities that are to be pursued with other individuals. So, this method needs to gather information ( data) to get the relationship between activities, travel and spending work time and space time to evaluate whether there are how many families who have real needs to spend time to go to travel in the year.

3.7 What is trip based versus activity based approaches?

What is trip based versus activity based approaches? The fundamental difference between the trip-based and activity based approaches is that the former approach directly focuses on trips without explicit recognition of the motivation or reason for the trips and travel. The activity based approach , on the other hand, views travel as a demand derived from the need to pursue travel activities. So, it is better understand the individual or family behavior basis for individual or family travelling decision regarding participation in travelling activities in certain places or cities or countries at given times and hence the resulting travel needs. This behavioral basis includes all the factors that influence the why, how, when and where of performed activities and resulting individuals and household, the cultural/ social norms of the community and the travel surrounding environment.

Another difference between the two approaches is in the way travel is represented. The trip based approach represents travel as a collection of trips. Each trip is considered as independent of other trips, without considering the inter-relationship in the choice attributes , such as time, destination and mode of different trips. As tours are chains of trips beginning and ending at a same location , say home or work. The tour based representation helps maintain the consistency across and capture the interdependency and consistency of the modeled choice attributed among the trips of the same tour.

In addition to the tour based representation of travel, the activity based approach focuses on sequences or patterns of activity participation and travel behavior, using the whole day or longer periods of time is the unit of analysis. Such as approach can address travel demand management issues through an examination of how people modify their activity participation, for example, will individuals substitute more out-of-home activities for in home activities in the evening of who arrived early form work due-to a work schedule change?

The major difference between trip based and the activity based approaches is in the way, the time dimension of activities and travel is considered. In the trip based approach, time is reduced to being simply a cost making a trip and a day's viewed as a combination, defined peak and off peak time periods. On the other hand, activity based approach views individuals' activity travel patterns are a result of their time use decisions with a continuous time domain. As individuals have 24 hours in a day or multiples of 24 hours for longer periods of time and decide how to use that travel among or allocate that time to activities and travel and with who, subject to their socio-demographic, transportation system and other and scheduling of trips. So, determining the impact of travel demand management policies on time use behavior is an important step to assessing the impact of such policies on individual travel behavior. The final major difference between this two approaches relates to the level of aggregation. In the trip based approach, most aspect of travel, e.g. number of trips etc. are analyzed at an aggregate level.

Consequently, trip based methods accommodate the effect of socio-demographic attributes of households and individuals in a very limited fashion, which limits the activity of the method to evaluate travel impacts of long term socio-demographic characteristics of the individuals who actually make the activity travel choices and the travel service characteristics of the surrounding environment. So, the activity based models are better equipped to forecast the longer term changes in travel demand in response composition and the travel environment of urban areas. Also, using activity based models, the impact of policies can be assessed by predicting individual level behavioral responses instead of employing trip based statistical averages that are aggregated over defined demographic segments.

3.8 Can apply (AI) big data gathering method predict senior age will be main travelling target?

In the past, Germany government had established tourism survey analysis to analyze survey data in order to arrive at reliable conclusions on future trends in travel behavior. To aim to find how demographic change will influence the tourism market and how the industry can adapt to those changes. The travel analysis provided data on tourism consumer behavior, including attitudes, motives and intentions. Since, 1970 year, it is based on a random sample, representative for the population in private households aged 14 years or older. Then, a continuous high scientific standard combined with a national and international users makes the travel analysis a useful tool and reliable source for tourism industry and policy decisions. It aimed to gather statistical data. e.g. on the age structure and on demographic trends, quantitative and qualitative analysis with time series data from the travel analysis. It shows e.g. not only the future volume , quite different from today's seniors, or how who will travel of family holidays will change, e.g. single parents of low, but grandparents of growing significance for tourism.

Demographic change is said to be one of the important drivers for new trends in consumer traveling change behavior in most European countries ( e.g. Lind 2001). Because the growing number of senior citizens in the European Union and other industrialized countries, such as the USA and Japan, looks to become one of the major marketing challenges for the tourism industry. United Nations statistics predict that the share of people being 60 age or older will grow dramatically in the coming future, and is expected to rise from 10 percent of the world population in 2000 year to more than 20 percent in 2050 year ( United Nations Population Division, 2001). From its statistic, some data showed that travel propensity increased throughout life until the age of about 50 years of age and was then kept stable until very late in life 75 age. The most important results is that the travel propensity when getting older is not going down between 65 and 75 age of course, the overall development of this variable is influenced by a lot of other factors which are responsible for quite a variation over time. It is now possible to suggest that the general pattern of travel propensity is one of the key indicators for holiday life cycle travel behavior, includes three stages. The growth stage tends to increase from early adult hood until 45 age old or when reaching some 80%. The next stage is stabilization from the ages of around 50 age, until 75 age old, starting with a lower increase. Finally, the decrease stage is a slight decrease occurs once people reach the more advanced age of 75 age to 85 age old ( Lohmann & Danielsson 2001).

So, it seems Germany government tourism prediction to future travelers‘ behavior indicated these findings, such as on how future senior generations will travel, who had used survey data to examine the patterns of travel behavior of a generation getting older and applied the findings to draw conclusions on the future. Also, it predicted that on the future of family trips, family segmentation will be the travel behavior patterns in the future. These findings together with the statistical data on demographic change allowed for a better understanding of the coming tends in family holidays. It's aim developed in consumer behavior related to demographic change and predicted what will happen future of tourism one had to consider other influences and drivers as well, for example, trends on the supply side. e.g. low cost airlines or in travelling consumption behavior in general whether how the past may provide a key to predict travel patterns of senior citizens to the future.

Given the projected growth of the senior citizens market, designing specific marketing strategies to meet the prospective needs of elderly tourists will become increasingly important. It has been an implicit assumption that it will be a close relationship between the travel behavior of today's senior citizens and the those of future ones. The growing number of senior citizens in the world. e.g. China, Hong Kong, Japan, USA etc. countries. Global senior citizen tourism market will be based solely on demographic predictions about the future of the population's age structure. However, many of these seniors won't only live longer but will be fitter and more active until later in life. Many of the will also have plenty in life. Many of them will also have plenty of time and money to spend on travel. So, will these new seniors behave like today's senior citizens? Will they adopt the same travel behavior as the previous generation or become a new market of oldies for the leisure and tourism industry? However, to determine the actual number of senior citizens who will be travelling and to sought to evaluate and specify certain difficult to predict the actual numbers of senior citizen to any country. However, they can be based on the implicit assumption that there is a close relationship between the travel behavior of past, present and future seniors. But is this a valid assumption? As the revise- analysis travel analysis survey, which was conducted in Germany every year, offered some interesting data possibilities. It was designed to monitor the holiday travel behavior, opinions and attitudes of Germans and has been carried out since 1970 year, questions in the questionnaire. Data are based on face to face interviews, with a representative sample of more than 7,500 respondents, the interviews being carried out in January each year. All results refer to the average for the defined generated, which ranges generally over ten years. The group of people then at the age of 60 to 69 age is described. This corresponds to the same generation ten years ago, when they had an age of 50 to 59 age. When this methodological approach is not necessarily very sophisticated, it does have the important advantages of being cost effective.

3.9 IS (AI) big data gathering method a better psychological method to compare human marketing research method predict travel behavioral consumption?

On the psychological view point, I think individual traveler's character will have those kind of personal characteristics. First, simplicity searchers value above everything ease not transparency in their travel planning and holiday making, and are willing to avoid having to go through extensive research.

Second, cultural purists use their travel as an opportunity to immerse themselves in an unfamiliar looking to break themselves entirely from their home lives and engage. Sincerely with a different way of living. Third, social capital seekers understand that to be well travelled is a personal quality, and their choices are shaped by their desire to take maximum of social reward from their travel. They will exploit the potential of digital media to enrich and inform their experiences, and structure their adventures always keeping in mind they are being watched by online audiences. Finally, reward hunters seek a return on the investment who make in their busy , high-achieving lives. Linked in part to the growing trend of wellness, including both physical and mental self-improvement who seek truly extraordinary and often indulgent or luxurious' must have experiences.

Why needs to know the personal character of individual traveler's characteristics? Because if travel agents could feel which kinds of individual traveler's character, then who can predict which kind of travel package to design to them more easily. For example, how to determine future travel behavior from past travel experience and perceptions of risk and safety? We need to concern that the influences of past international travel experience, types of risk associated with international travel and the overall degree of safety feeling during international travel on individual's travelling experiences likelihood of travelling to various geographic regions on their next international vacation trip or avoidance of those regions, due to perceived risk. Because individual traveler's experience of safety risk degree to the countries, it will influence who chooses to go to the countries/ country to travel again.

Why travelers avoid certain destinations are as relevant decision making as why who choose to go to the country(countries) to travel. Perceptions of risk and safety and travel experiences are likely to influence travel decisions; efforts to predict future travel behavior can benefit to individual tourist's decision making.

As Weber & Bottorn (1989) defined risky decision is as "choices among alternatives that can be described by probability distributions over possible outcomes" (p.114). Some psychologists judge subjective perceptions of physical reality, i.e. image of a particular tourist destination, whereas value judgement refers to the way individual rank destinations according to whose attributes. i.e. attractiveness, safety, risk etc. factors to form on overall image. So, if the individual traveler had unhappy and worried and unsafe experiences to go to where the place(country) to travel during whose vacation time before. Then, this negative travel experience will influence who is afraid to go to the place ( country) to travel again. Risk of place, country, destination or region means the danger is relatively high to the place, i.e. increasing in airplane accidents, crime or terrorist activity targeting citizens of potential traveler's nationality or the probability of occurrence is great , i.e. recent occurrences involving travel regions/ destinations under consideration or effective actions to control consequences exist. i.e. selecting safe regions and destinations, taking extra precautions when traveling to risky destinations. These risk factors will influence the individual traveler who chooscs to cancel travel plan to go to the country again.

Another interesting research, how to predict behavioral intention of choosing a travel destination, which has focus of tourism research for years, but the complex decision making process leading to the choice of a travel destination has not been well researched. The planned behavior model using its core constructs, attitude, subjective norm and perceived behavioral control, with the addition of the past behavioral variable on behavioral intention of choosing a travel destination.

Understanding why people travel and what factors influence their behavioral intention of choosing a travel destination is beneficial to tourism planning and marketing. Understanding travel motivation is the push and pull model. The idea of the push and pull model is the decomposition of an individual's choice of a travel destination into two forces. The first force is the push factor that pushes an individual away home and attempts to develop a general desire to go somewhere else, without specifying where that may be. The second force is the pull factor, that pulls on individual toward a destination, due to a region specific travel location or perceived attractiveness of a destination. The respective push and pull factors illustrate that people travel because who are pushed by their internal motives and pulled by external forces of a destination. Nevertheless, how push and pull factors guide people's attitude and how these attributes lead to behavioral intentions of choosing a travel destination have rarely been investigated. The decision making process leading to the choice of a travel destination is a very complex process. The planned behavior model is as a research framework to predict the behavioral intention of choosing a travel destination. The model based on the three constructs of attitude, subjective norm, and perceived behavioral control ( Fishbein & Ajzen, 1975).

In conclusion, the factors can influence travelers who decide to choose to travel the country, which include personal safety was perceived to the highest motivation factors among the important factors which include, scenic beauty, cultural interests, friendliness of local people, price of trip, services in hotels and restaurants, quality and variety of food and shopping facilities and services. The factors include both push and pull. Push factors include knowledge, prestige, and enhancement of human relationship etc., whereas, the most significant pull factors include high technologic image, expenditure and accessibility etc. For example, Japanese travelers visiting Hong Kong. Push factors are such as exploration dream fulfillment and pull factors are such as benefits sought, attractions and good climate city. It will be the factor of future travel patterns and motivations of sub-cultural and ethic groups for Japanese choice to go to Hong Kong travelling.

3.10 How can apply (AI) digital channel ( big data gathering method) predict travelling consumer behaviors?

(AI) big data digital channel can be applied to help travelling businesses to evaluate whether how much the e-ticket price and travelling package price is the most attractive or reasonable to persuade travelling consumers feel it is the most reasonable price to choose to buy the airline's e-tickets or the travel agent's travelling package product from internet channel . It helps travelling consumers to feel which airlines or travelling agents which ought change their e-ticket and/or travelling package price to let travelling consumers to choose to buy the airline e-ticket or the travelling agent's travelling package products from internet channel. It can be applied to predict whether how many travelling consumer numbers can be increased or decreased when the airline e-ticket price is variable or the travelling agent travelling package price is variable . It aims to give opinions to help any online airlines or travelling agents to judge whether which e-ticket or travelling package price is the most reasonable to let travelling consumers to accept to choose to buy which airline's e-tickets or traveling agent's package products more attractive.

Thus, (AI) e-ticket or e-travelling package price measurement technology can be preference to be applied online communication ecommerce and mobile phone internet platform aspect. As traveling businesses can enter their past e-ticket or travelling package prices data and past travelling customer number data into computer or mobile. Then, (AI) price measurement technology can gather these data to analyze these e-ticket or travelling package product prices and past travelling customer number to compare their e-ticket and/or travelling package prices variable changing range level to find their e-ticket and /or travelling package price variable difference to measure to make conclusion about every travelling package or/and e-ticket product's price variable changing will influence how many travelling customer number increase or decrease changing to choose to sell their different kinds of travelling package or e-ticket products more accurate. Then, (AI) price measurement software will help them to analyze all past e-ticket and/or travelling package price variable changing data to compare whether which e-ticket and/or travelling package price range can let travelling customers to feel it is more reasonable and attractive to influence them to choose to buy their e-ticket or travelling package product among different airlines and travel agent choices. Because any e-ticket or travelling package product's price is one important factor to influence travelling consumers to choose to buy the airline's e-tickets or travelling agent's travelling package products.

For example, Amazon publish has applied (AI) price measurement technology to help authors to decide how much every different topic of e-book or paper book price, it can attract the largest number of readers to buy. Any one author only needs to type whose book name to Amazon publish author himself/herself Amazon website. Amazon publish (AI) price measurement learning machine will help them to auto-calculate and judge how much e-book or paper book price is the most attractive and the most reasonable in order to increase reader number to buy their e-books or paper books to read. So, (AI) online price measurement machine will gather past similar book names and past every similar book readers' reading times and the number of readers to give opinions to let every author to judge whether his/her very new e-book or paper book ought charge how much price to the e-book or paper book which can attract many readers to choose to buy. Although, it is not ensure that the e-book or paper book price must let readers to feel it is the most reasonable price to choose to buy in reader's view point. However, it has other factors to influence readers' choice to buy the e-book or paper book, e.g. whether the book content is attractive to public, the author's familiarity, the book's page is enough or not to satisfy readers to read etc. factors. But, instead of all these extra factors to influence readers to choose to buy the book to read. (AI) price measurement learning machine can real give opinions to every author to let them to judge the e-book or paper book different price range whether is too high to influence readers to choose to buy to read or tool low to influence readers feel it is possible poor content book to compare other similar content books. Thus, (AI) price measurement machine can help authors to predict every reader's reading behaviors or reading experience and reading habit from online channel in short time easily. The author only enter the book name to let Amazon publish price measurement machine to check, it will follow past reader's reading habit and reading experience to judge whether the similar all book topic sale record to judge how much price is the reasonable price to attract many readers to buy the book.

Hence, (AI) can be applied to digital channel to help travelling businesses to predict travelling consumer behavior in the future. In the future, mobile/ smartphone, laptop, desktop will be most frequent used ecommerce channels to develop online business. So, (AI) can be also applied to these platforms to gather data to make analysis to help travelling businesses to predict travelling consumer purchase behaviors popularly. Due to , ecommerce is popular to global, so digital online and instore channels can be one good channel to let (AI) learning machine to make platform to gather past every online travelling consumer purchase ( buying) experience data to help travelling businesses to build airline or travelling agent brand personality and having a responsible, positive impact on society.

To apply (AI) learning machine technology to understand travelling customer online purchase behavior, it will raise business e-commerce successful chance: For example, (AI) learning machine can help travelling businesses to gather data to analyze to determine whether short-term or long-term signals in the online travelling consumer behavior that indicate higher purchase intents to let every online travelling business to know. (AI) learning machine can find that online users with long-term purchasing intent tend to save and click through on more content.

However, as online travelling users approach the time of purchase their activity becomes more topically focused and actions shift from saves to searches from online travelling consumption channel. Then, (AI) learning machine will further find that the brand airline and/or travelling agent purchase signals in online travelling consumption behavior can exist weakness before an online travelling purchase is made and can also be traced across different online travelling purchase categories. Finally, (AI) learning machine synthesize these insights in predictive models of online travelling user purchasing intent to the brand of airline or/and travelling agent travelling package product. Taken together, it's work identifies a set of general principles and signals that can be used to model online travelling user e-ticket and/or travelling package purchasing intent across many online content discovery applications. Thus, (AI) learning machine can help online travelling businesses to gather any online travelling users' click online travelling behaviors data to judge whether there are how many online travelling users will choose to find their online travelling business websites to make final decisions to buy their travelling package or/and e-ticket products from online channels. Then, it will give opinions to help the online travelling businesses to let it to judge whether what are the important website factors will help its online travelling business to attract many online travelling consumers, e.g. designing unattractive travelling website issue, online unattractive scene photos issue, unclear website travelling photo color issue, unclear website travelling advertisement message, contents and words impressions issue, lacking image movement frequent attractive seeing issue etc. different website factors. Thus, online digital channel will be one good choice to apply (AI) learning machine to help travelling businesses to predict travelling consumer behaviors.

Thus, (AI) big data technology can also assist travelling consumers to gather different manufacturers' data to compare what their advantages and disadvantages of their travelling package products are. Then, travelling consumers can make comparison to choose which airline or travelling agent is the suitable to whom to buy e-ticket or pre-booking travelling package in online travelling consumption market.

.

Thus, I believe that artificial intelligent "big data" gathering method can be suggested to be applied to attempt to predict travelling consumer behavioral changes in global online travelling business environment, the reasons are as below:

On the travelling consumer's beneficial hand, travelling consumers can apply this (AI) big data gathering method to attempt to gather any global airline e-tickets and/or travelling agent's package product data to be analyzed by this artificial intelligent learning system to compare human general marketing research method, e.g. survey, questionnaire, marketing plan etc. different human judgement methods to predict traveler consumption behavioral change model. Then, it analyzed all the different data to compare what are the range of the most reasonable e-ticket and/ or travelling package online purchase history and sale in order to make more accurate prediction to future traveler change traveling consumption behavioral model in next month, or next half year or next year short term period traveling consumption change prediction. Thus, it seems that future AI tool can be attempted to apply to predict any industries price behavior, e.g. deciding what level of price is the attractive level to attract consumer in these industries, e.g. fuel, education, tourism, health, entertainment, etc. different product purchase. It can give more absolute price suggestion to any merchants to set their price change predict in order to increase many customer numbers to buy their products in every year, or every quarter every month, or month week, even every day etc. different sale period.

Reference

Backman and others “motivation is conceptually viewed as “ a state of need, a condition that services as a driving force to display different kind of behavior toward certain types of activities, developing preferences, arriving at some expected satisfactory outcome.”, 1995, p.15.

Fishbein & Ajzen, “The model based on the three constructs of attitude, subjective norm, and perceived behavioral control”. 1975.

Hsu et al. “A tourist behavior model has been developed, called the expectation, motivation and attitude “ ( EMA) model ,2010.

ICT,WWW . “Switzerland has one of the highest population-to-computer ratio in Europe.” Switzerland, 2005.

Jorea Ministry of Environment, “ For South Korea environmental attitude is a major factor in decision making vis-a-vis the consumption of " green" food and services”, Korea, 2015.

Korea Ministry Of Environment. Public Organizations spend 2.2 Trillon Korean Won To

Purchase green Products in 2014; Ministry Of Environment: Sejoung, Korea, 2015.

Lind , Lohmann & Danielsson , United Nations Population Division, “Demographic change is said to be one of the important drivers for new trends in consumer traveling change behavior in most European countries”. 2001.

Mayne, Lonnie. " Evolve of die in the age of the consumer". Entrepreneur, N.P. , 16 Apr. 2014. web of Oct. 2016.

Lee, D.; Kim, M. ; Lee, J. adoption of green electricity policies: Investigating the role of environmental attitudes via big data-driven search-queries. Energy policy 2016. 90, 187-201.

Lee, Terrence, " Tech in Asia-connecting Asia's startup system " Tech. in Asia- connecting Asia's startup ecosystem, N.p.,4 July 2016.

Weber & Bottorn "risky decision is as choices among alternatives that can be described by probability distributions over possible outcomes" , 1989, p.114.

# AI predicts energy consumption behavior

CHAPTER SEVEN

# AI predicts householder electricity energy consumption behavior

Can apply artificial intelligence measure the house quality or occupancy how influences the householder electricity energy consumption or useful activities to be more or less? In general, house owners have both intentions for whose property. One intention is living the house by householder himself or herself or householder with families themselves. Another intention is that renting to others to receive rent income ( landlord). So, in the housing market, the housing consumer includes either the property owner intents to rent to others to live for rent income aim or the property buyers intents to buy the house to be house owner to live. Does these both different property purchase intentions, which will influence the householder's attitude to use electricity energy consumption desire to be more or less, due to the householder's demand to whose house quality factor influence? This is one interesting question concerns the householder electricity energy consumption desire change to the householder, due to house's investment or house's living intention influences to house quality factor.

How does house quality factor influence to householder electricity energy consumption desire to be more or less? Has it relationship between house quality and house investment or living intention to cause house quality demand to influence the householder electricity energy consumption desire change or demand to be more or less? Has it relationship between regional housing market living or rent investment intentions, housing quality and electricity energy consumption more or less desire? I suppose that the determinants of the residential electricity energy demand form space-heating and cooking, due to the property quality demand influence and the householder's living or rent investment intention influence both, which will influence the householder's electricity energy consumption or useful behavior when he/she/they is/are living in the house.

I argue that rent properties are not only consumer goods, but it also constitute financial market assets. It is therefore reasonable to assume that rational ( rent income investment intention) investors choose to raise housing quality ( e.g. thermal insulation technological installing at home, heating or cooling technology or artificial intelligent window, lighting, door opening or closing) in order to attract many people choose to rent whose house to live. The householder's aim is to achieve an acceptable return on investment ( ROI) or raising rent income aim when he/she rents whose house to anyone, it is easy to attract many people to choose to pay higher rent his/her house to live in the property rent market. Moreover, the another important factor is that rents and future house sale prices of properties differ regionally ( or even locally), and largely depend on housing market fundamentals, such as either the house living buyer's income levels or the house rent buyer's income levels, vacancy rates, and/or householder investor's expectations.

Thus, if the householder expects to rent whose house and raises rent to attract many people choose to rent whose house to live, who will attempt to install many new technology in order to satisfy their high quality of life need when they can pay higher rent to rent to choose to rent whose houses to live. Their aim only achieves to raise housing quality, but any new technology will lead to increase electricity energy consumption or use in the house.

Hence, any high quality of houses will influence the householder to use or consume more electricity energy at home. It means that the householder will choose to consume or use more electricity energy at home, if he/she or the family householder demands to live more comfortable house and he/she/they can have high quality of living life at home. This comfortable living demand to the householder ( property renter or property buyer) view point can explain why the better quality of house factor will influence the electricity energy consumption desire to the householder also to be more daily.

I shall indicate one home electricity energy consumption experiment, it indicated that utilizing aggregate data on regional space-heating energy consumption form over 300,000 apartment buildings in 97 German planning regions. The study applies structural equation modelling to estimate the influence of housing market fundamentals on the level of housing quality, and subsequently on regional electricity energy consumption. Consequently, it suggests that housing market fundamental explain regional differences in the housing quality.

In particular, findings show that the level of per capita income, investor' expectations about future housing market development as well as vacancy all explain regional differences in housing quality has a significant impact on electricity energy consumption.

In the way, this experiment can indicate evidence that regional housing market fundamental have a substantial influence on regional levels of housing quality and energy consumption desires to the German regional householders.

This Germany regional householder experiment found important implications for high or low housing quality of the regional property building and householders either property living or property rent intention of comfortable living feeling need factor which will influence the regional property householder electricity energy consumption desire to be raised or reduced. These factors will influence the consequence of electricity energy demand to be increased or decreased needs every day for the regional householder as well as the country's electricity energy supplier(s) can gather the regional properties whether they are high or low quality to predict the regional properties householders' electricity energy consumption supply budget more accurate. It implies that an important determinant of residential housing quality will have possible to influence electricity energy demand to be more or less for long term. IN particular, this Germany regional residential experiment can explain and find an important role in formulating assumptions about the quality factor has chance to influence the regional residential future levels of electricity energy efficiency and consumption in the country. Hence, housing developments and electricity energy firms can follow this regional residential housing quality factor to evaluate whether the regional housing market is the corresponding investment patterns as well as the energy researchers can follow the regional residential housing quality whether it is high or low housing quality factor to evaluate the more accurate models of regional electricity energy demand to any regional residential householders' houses in the country.

In consumer behavioral view point, it explains that if the country government expected many householders feel to need to spend much electricity energy or have much electricity energy useful demand or desire at home. The country government ought to encourage the country residential property or house developers choose to build many houses which have technological product installed to satisfy the regional householders' residential comfortable living need when they choose the regions to build the high quality houses to let them to live. Then, the regional householders will be influenced to consume or use much electricity energy at homes, due to they feel that they are living at high quality and comfortable and high building technological installed apartments in the country's regions. Then, the country's government and electricity energy provider(s) may be raise much electricity energy efficiency and supply and profit , due to the regional residential householders' electricity energy consumption or useful desire need is therefore influenced to be more by the regional high quality of residential houses factor. So, the regional high quality of residential house factor will have relationship to the regional electricity energy consumption and efficiency to the regional householder's houses.

Otherwise, if the country government felt electricity energy is shortage, it ought encourage the property developers build many low quality and low building technological houses to let householders to live themselves or rent to others to live in the country's different regional residential development market. Due to the low quality of properties and low technological installed to properties factor which will influence any these different regional residential householders to choose method to solve shortage of electricity energy challenge to the country.

In conclusion, to apply consumer behavioral economic theory to property development market, if property quality factor can really influence the householder's electricity energy consumption desire to be used more or less at home daily. The country's property developers can apply this factor to predict property consumer individual property buying consumption behaviors more accurate. For example, if the US property developer planned to build low quality and low technological design buildings and lesser comfortable residential houses in the region in US. Then, its residential householder target will be trended the less acceptable of electricity energy consumption property buyers to choose to buy these regional properties to live in the US region because they can only accept to spend less electricity energy to use when they are living in the houses in order to save money daily. SO, the low quality , less comfortable and low technological installed design residential houses will satisfy their living needs. Otherwise, if the US property developer planned to build high quality and high technological installed design buildings and more comfortable residential houses in the region in US. Then, its residential householder target will be trended to the more acceptable of electricity energy consumption property buyers to choose to buy these regional properties to live in the US region because they can accept to spend more electricity energy to use when they are living in the houses in order to improve their living of quality. So, they wont's consider to spend more expenditure to use electricity energy for any technological products are installed in their properties in order to satisfy their comfortable living needs at their homes every day.

Consequently, property developers can attempt to gather marketing research concerns whether how many people who accept to use more electricity energy or use less electricity energy in order to predict they ought build how many high quality or low quality houses number in different regions more accurate in themselves countries or overseas countries property development market.

1.1 AI measurement environment factors influence householer electricity energy consumption activities

Can artificial intelligence measure whether what environment factors influence householder electricity energy consumption activities? Has environment factor relationship to influence householder electricity energy consumption behaviors? Socially, householder electricity energy consumption provides us with sources of living satisfaction , but if any sudden environment factor changes, whether it will influence householder consume or use more or less electricity energy decision at home. However, I assume householder electricity energy consumption will have a considerable proportion of the environmental impacts be influenced by our way of life and our economic decision of electricity energy consumption behavior.

What different environmental factors will influence householder electricity energy consumption decision? The external environmental factors include, for example, the country's electricity firms or government changes to electricity energy regulations, electricity energy production technologies change and business practices and government policies changing etc.

different external environmental factors will influence any country's electricity energy consumption to householders' consumption desire to be more or less. It will also require changes to influence the householders to consume which kinds of electric products which are needed to be used in different electricity energy natural manufacturing resources.

Why does these external environmental factors impact householders' any behaviors to influence them to concern to use more or less electricity energy power or which kinds of electricity energy products choice at homes. How any why environmental factors impact will influence householder activities at home, such as electricity energy consumption and choice? What are the key components of external environmental factors influence householders' electricity energy consumption behaviors. I shall explain as below:

Firstly, we need to know whether what external environments are which can influence why and how householders need to change their activities to choose more or less or which kinds of energy power to be provided to them to use at home. Who is householder? Householder is an individual, family, or group of individuals living together as unit in a home. Consumption of electricity energy at home may be cooking food needs, needing have colder feeling to turn on fan or air condition at home in summer or needing have warm feeling to turn on heater at home in winter, watching television programs or listening music , playing computer games or used computers activities , reading activities and applying artificial intelligent technological tools to help householders to open or close homes' windows, doors etc. different home equipment which need to use electricity energy provisions. SO, their home activities need to turn on lighting electric tools , televisions, music machines, radios etc. different equipment which need to use electricity energy provision at home. SO, the purpose of householder consumption means consumption by individuals living in a household and it includes consumption both in and outside the home. Why does environmental impacts link to householders' electricity energy consumption? I shall focus on discussing of greenhouse gases ( GHGS) energy product how any why it can influenced to householders to use.

The environmental impacts will influence this kind of greenhouse gases ( GHGS) energy in the product lifecycle or delivery of the service to link the householder's energy consumption at home such as these several aspects:

Extraction and greenhouse gases production ( supply number), physical distribution ( delivery far long or close near short distance between the greenhouse gases manufacturing factory and the greenhouse gases supplier), resources consumed by marketing and retail activities ( householder's needs to use the quantity of the greenhouse gases energy product), the greenhouse gases consumers search and purchasing activities( e.g. travel to shops, internet purchasing channel, , finding the which kinds of greenhouse gases products from internet, magazines, newspapers, radio advertisements etc. different medias,) , post-use greenhouse gases energy disposal ( resale, reused or rubbish). The householder's physical behavioral impact environmental factor will influence how and why he/she chooses to consume greenhouse gases energy daily , e.g. impacts of a housing development, or a wind –farm that supplies greenhouse gases with power. So, the householder's greenhouse gases energy consumption behavior which will depend upon individual personal and subjective perspectives and value.

So, the householder's useful behavior or attitude of greenhouse gases energy product which will influence how he/she/ the family use or consume greenhouse gases energy, such as the householder individual environmental protection attitude which can impact how he/she/the family spends the quantity of greenhouse gases energy every day at home, if the householder does not expect our air or water or land is polluted , due to extraction of any natural gas resources to be manufactured any kinds of greenhouse gases products. Then, this environmental pollution issue will influence some householders choose to reduce to use more quantity of greenhouse gases products every day. Another environmental factors include the bio relates the ( unsustainable ) use of resources to avoid wasting much greenhouse gases energy to cause greenhouse gases energy supply shortage, avoiding the cause negative impacts of quality life , e.g. noise causing when the extraction of any natural resource from lands to the householder's house is near to the natural resource extraction land and health impacts, e.g. when the greenhouse gas householder user who often use the kind of greenhouse gas product when it is used to cook or heat any equipment to cause they to breathe dirty air at home often. These impacts can be measured in different ways include: monetary costs or loss, physical quantities of resources used or waste or pollution produced and the burden the greenhouse gases energy place on environmental resources. All of these external environment factors will impact the householder individual attitude or behavior how to use or consume greenhouse gases energy product at home.

All these environmental factors concern householder greenhouse gases energy consumer individual consumption attitude is influenced by environment pollution, greenhouse resource supply shortage challenge, greenhouse gases influence the householder's negative quality of life, negative health impacts, noise, waste money , raising economic cost to the householder which will impact whether how the householder choose to use the quantity of greenhouse gases product or the kinds of greenhouse gases products or other kinds of electricity energy products.

However, these are other external environmental factors which can impact how the householder decides to use greenhouse gas product at home. They include: the changes of energy regulation, e.g. the country government has quota number implementation to prohibit to import above the limited quantities of any kinds of greenhouse gas products to any countries. So, when the greenhouse gas energy supplying quantity is decreased, but if the country has may householders who need to buy different kinds of greenhouse gases products to be used at home. Then, the different kinds of import greenhouse gases energy products prices will be raised in possible, due to demand is more than supply in the country's greenhouse gas energy product market. Consequently, if the greenhouse gas energy price us risen above the general social acceptable level to the home greenhouse gas energy product householder consumers. Finally, it will influence them to choose to buy other kinds of gas energy products to replace the greenhouse gas energy product to use at home.

Another side, if the country's greenhouse gas energy manufacturing supplier sudden changes its greenhouse gas energy production technologies to choose to concentrate on manufacturing other kinds of energy products. Then, the greenhouse gas energy supply quantities will be only decreased, even future one day , it will cause greenhouse gas supply shortage challenge to let the country's home greenhouse gas householder consumers who can not buy enough quantity of any kinds of greenhouse gas energy products to satisfy their electricity needs at home every day. Consequently, when future on day , the country greenhouse gas energy manufacturer has none any quantity of greenhouse energy products to supply to the country's greenhouse energy householders to use at home. The, they must only choose other kinds of new energy products to replace the traditional useful greenhouse gas energy products to be used at homes.

In conclusions, these non-controlled external environmental factors can impact and influence the country's every householder consumer individual attitude or consumption behavioral change to how any why the country's householders either choose to buy much or less quantity of greenhouse gas products to use at home.

1.2 AI measurement house space occupancy
and building characteristics how influence on householder electricity energy use

Can artificial intelligence mesaure how much space occupancy is the suitable size as well as what the most suitable building characteristics influence each householder electricity energy useful activities or behavious? In general, society believes large space size occupancy house building characteristics factor which will influence householder use more energy at home, e.g. in summer, when the householder is living at the large space size occupancy house, who ought turn on all air conditions or fans at sleeping rooms or eating room or studying room. So, if the householder's house has two to three or more sleeping rooms. Then, he / she needs to buy more air conditions or fans in order to let all rooms‘ temperature to be fallen down to let he /she feel more cool comfortable feeling when the temperature is above 30 degree or more extreme hot in summer weather. Otherwise, when the temperature is low, e.g. between 0 degree to 10 degree or below 0 degree in winter weather. When the householder is living in one large space size occupancy appartment, which has thee to five sleeping rooms , even more and two studying rooms and one eating room, even more as well as every room has one heater. Then, he / she must turn on all heaters to let who to feel warm feeling when he / she is staying in the house. It brings these interesting questions.

Will large or small size space occupancy housing characteristics influence any householder often turn on heater or air condition or fan in whole house space occupancy area in order to the householder feels warmer or cooler feeling when he /she is staying in the house?

Has any space occupancy housing characteristics relationship to influence any householder to turn on heater or air condition or fan in whole house space occupancy area in order to the householder feels warmer or cooler feeling when he /she is staying in the house?

Does it bring positive relationship between turning on long time fan or air condition or heater and the house occupancy space characteristics is large or small size?

I shall attempt to give psychological evidences to explain the householder's house space occupancy area large or small size factor whether it can influence the householder choose to do long time or short time turning on heater or air condition or fan behavior in order to let he/she/the family to feel more cooler or warmer comfortable feeling when he/she/the family is staying in the house in summer or winter weather.

Does the house occupancy space size characteristics factor is the only one or important factor to influence the householder choose to turn on long or short time fan or air condition or heater in the house to let him/her/ the family to feel more cooler or warmer comfortable feeling in summer or winter weather?

I feel that it is not exact right , due to the householder's house space occupancy size whether it is large or small characteristics to influence the householder choose to turn on long time or short time fan or air condition or heater time to let him /her/ the family to feel more cooler or warmer when he / she / the family is staying at home in summer or winter weather. The reason is because that the lifestyle of living quality need is different between developed countries and developing countries. The lifestyle of living quality factor will change the country's householder's expectation about the quality of living life. For example, for Africa, Korea, China , Japan, Hong Kong etc. developing countries. On the lifestyle of living quality need to these developing countries' householders aspect, that will cause a high environmental burden when they need to often turn on air conditions to satisfy more cooler feeling when they are staying at homes in summer or they need often to turn on heaters to satisfy more warmer feeling when they are staying at home in winter. Due to if their houses are large size space occupancy characteristics and they have more than at least two sleeping rooms and studying rooms and eating rooms and toilets number. Then, these householders who are developing countries' large space occupancy size characteristics houses, they won't like often turn on heaters long time to keep more warmer in their indoor whole space area in winter or they won't like often turn on air conditions or fans long time to keep more cooler in the their indoor whole space area house environment in summer .

The reason is possible because that the developing countries' householder chooses often to turn on their heaters or air conditions or fans long time in their houses when they are staying long time in their houses and their houses space occupancy sizes are very large, it will bring the electricity energy to be used more to these developing countries' householders' large space occupancy size characteristic houses. It means that the electricity fee will be also increased due to they often turn on heaters or air conditions or fans long time to keep their indoor temperature to be more cooler in summer or more warmer in winter. So, it seems that the developing countries‘ householders are living in the house whose space occupancy have very large size characteristics and more than two rooms house characteristics in the developing countries as above. Then, they won't often choose to turn on heaters or air conditions or fans long time to keep more cooler or warmer feeling in their house whole indoor space occupancy environment when they are often staying at home long time.

Their lifestyle of living comfortable feeling are lesser than the developed countries householders. Consequently, their lesser cooling or warming comfortable demand of living lifestyle factor will change their attitudes to use air conditions or fans or heaters turning on time in order to limit heaters or air conditions or fans turning on time to be shorter than the developed countries houeholders' heaters or air conditions or fans turning on time at homes. Due to the long time turnong on air conditions, fans , heaters at the developing countries‘ householders' homes, it will cause to spend much electricity energy to lead electricity fee charges to be raised to the developing countries‘ householders ' homes when they are often staying at homes in summer or winter weather. Hence, the house space occupancy large size characteristics ought not influence the developing countries householders choose to turn on air conditions , fans or heaters long time in order to let them to feel more cooler or warmer at homes in summer or winter weather.

So, the developing countries‘ house space occupancy large size characteristics householders won't be more acceptable to pay higher electricity energy fee when they are staying at homes at summer or winter weather. Due to they do not often choose to turn on heaters, air conditions or fans long time during they are staying at homes. Otherwise, the developed countries, e.g. UK, UK , France, Germany, Swiss, Singapore, Italy etc. countries. In general, these developed countries' householders‘ living lifestyle quality needs are higher than the developing countries. So, when the summer or winter weather is coming, if the temperature is extreme cold, e.g. below than 0 degree or it is extreme hot, e.g. higher than 30 degee. Then these developed countries' householders will easy accept to turn on air conditions or fans or heaters long time at home in order to keep their appartment in door temperature to be more cooler in extreme hot in door environment or more warmer in extreme cold in door environment when these developed countries‘ householders are often staying at homes long time at night after their day time working time or schooling time. Because these developed counties' householders' quality of living lifestyle needs or demands are higher than the developing countries' householders. So, they won't consider that they will pay more electricity fee , due to they often turn on air conditions, fans or heaters long time to let them to feel more comfortable in cooler or warmer indoor large size space occupancy environment. So, it seems that the electricity energy efficiency will be raised to the developed countries' householders who are living in the house space occupancy large size characteristics and they will be possible to pay more electricity fees during they are often staying at home in extreme hot summer or extreme cold winter weather.

In conclusion, due to the living lifestyle quality need ( demand) is different between the developed countries' householders and the developing countries' householders. It will influence the householders' long time or short time spending time on air conditions or fans or heaters indoor space occupancy size characteristics environment in order to achieve more cooler or more warmer feeling in their houses. Consequently, the long or short time of turning on air conditions, fans, heaters for the developed or developing countries householders' activities factor will be more influential to compare the house space occupancy large or small size characteristics factor to influence their cooler or warmer feeling in their houses. SO, the house indoor environment electricity energy consumption efficiency degree to the developing or developed countries' every householder house in summer or winter to the developing or developed householders in summer or winter weather , which is more influenced by the living lifestyle qualty factor to the either developed countries or developing countries householders. Hence, any developed or developing countries' electricity suppliers need to consider the building areas of property development market buyers their living style quality demands ( needs) whether their living style quality demands are higher or lesser than the other building areas of property development market, they ought not consider whether the building locations of the houses' space occupation sizes whether they are large or small sizes in order to evaluate the householders will live at the building areas of property development locations ,whose electricty energy spending efficiency more accurate.

CHAPTER EIGHT

# AI measurement how Social factor influences household renting behavioral choice

Can artificial intelligence find what social factor influences household renting behavior ? Can rent subsidies assistance social behavioral factor influence the country's householders' housing needs in society? Why and how will the impact of social information factor influence on willingness to householders to pay rent in the country? The housing rent decisions will be determined by objective factors as well as housing market renting decision will be depended by future housing consumption desire beneficial influence as well as it is relative to current housing investment and consumption. If the householder expected or predicted future housing choice to pay rent to live because he / she feels future housing price will fall down in possible.

Hence, if the country's government can provide rent subsidiaries assistance to its householders. Also is the country has many householders feel the housing price will be fall down within five years for example. Then, it will be possible encourage that many householders choose to pay rent to live within five years. So, they country government's tent subsidiaries assistance will encourage the country's renting house market to be grown up rapidly within five years. It means that the social expectancy factor will have indirect influence to impact household housing buyers or renters' housing consumption behaviors.

However, in addition, the above householder expected or predicted future housing choice social information factor influence , other social factor influences can also influence housing activity, they include such as : When conditions are favorable in terms of house price, property taxes, fiscal incentives etc. factors. These social factors will impact how many household property consumers how to make house renting or buying decisions. SO, property buyers and property sellers both will have incentives to wait for more information about the state of the housing market before deciding to buy or rent house to live. However, the property buyers and sellers have limited information about the real intentions of transacting parties, the probabilities of renting or buying property decision will have possible increase the riskiness of housing decision.

A behavioral housing model is constructed which allows that the country householders' housing decisions of owner occupies are the outcome of interactions between objective analysis and subjective behavioral factors ( social factors) of the householder himself / herself decision factor.

The social factors, such as objective process of social learning or / and the more subjective influence of social pressure, which will influence the householder's renting or buying house decision when he / she feels have living need in society. For social learning expectations, the impact of uncertainty and judgements about future benefits and costs of housing. So, every householder will usually gather information from social medias to attempt to predict when the country's houses prices will be fallen down or risen up. SO, it explains why some householders who have enough money to buy houses to live, but they choose to rent houses to live in themselves countries. Because their judgements let them to feel that there is possible to reduce housing price in society, e.g. after one year. Hence, this " after one year, the houses prices reducing information or message in society will influence the country's many householders choose to rent houses in the year. Because they feel that it is possible the society's houses prices will be decreased after this year. This " after one year housing prices reducing social information or message factor" will influence the country's many householders choose to rent houses to live in this year. They expect the social houses prices to be fallen down after one year, then they will choose to buy the falling down prices of houses to live.

After aspect of social influence to the country's householders to choose either house renting or buying behaviors it normative influence , such as housing renting or buying case, when an householder's renting or buying one house decision is affected by social media perception of information as well as the householder individual's attitude and response of others.

The householder individual's renting or buying one house's decision which will be influenced by the social media information influence. SO, it makes sense to judge that the crowd is thinking in the very short-term and moving housing renting or buying markets in the country.

The information that is important that other many people are prepared to pay rent to rent the houses in the country, when the householder ( either renting or buying house ) consumer thinks that housing market valuations of house assets are going to increase in value, even though the householder may privately think the future house sale prices are over-priced.

These external housing market valuations information may influence the householder ( either renting or buying house) consumer to make decision when he/she feels that it is the right time , he/she ought choose to buy one house or rent one house to live in the country. So, at this moment, he /she will prefer to pay long-term rent expenditure to rent one house to live, due to the housing market information indicates the future housing market valuation will decrease in order to influence he/she can pay lower housing price to compare to buy one house today. So, at this moment, it is the right time that the country can implement rent subsidiaries assistance to support the low income working age group to have effort to rent low quality of houses in society.

Consequently, it explains why social housing media factor and the country's government 's rent subsidiaries assistance implement will influence when the country's householder planners who choose to buy or rent houses to live in the country. For example, if the country's government encouraged people to rent houses to live , when the country's government predicts that the country's any houses prices are raising to above the society's houses purchase burden effort in society. Then, the country's government can provide rent subsidiaries assistance to excite or encourage many low income working age group people to choose to rent houses to live. Otherwise, when the country's government predicts that the country's houses prices will decrease in future day. The country government can choose not provide rent subsidiaries assistance to attract or encourage them to rent houses in society. Then, it will encourage many people to choose to buy low price houses to live.

In conclusion, this rent subsidiaries assistance strategy can help the country's housing market to achieve the more housing supply and demand balance between house renters and house buyers both. It is beneficial to develop the property renting and property purchase market development in the country.

2.1 artificial intelligence estimates how much rent subsidiaries is the most right level to attract householders to choose to rent houses in preference

Can artificial intelligence estimate how much rent subsidiaries is the most right level to attract householders to choose to rent houses in preference ? When one country provide rent subsidiaries assistance to support people to pay lesser rents in the country. Can this country housing benefit bring advantages for landlords in society? Although, in the country's society, it is possible to encourage many people to choose to rent houses in the country. It seems that the country's landlords will increase tenants number and they can earn more renting income in short term benefit. But, in long term , it will bring negative influences to landlords. I shall explain the reasons as below:

When, the country provides long term rent subsidiaries assistance to its people, e.g. five years rent subsidiaries assistance. Although, it can encourage many people ( tenants) choose to pay lesser rent expenditure into accommodation in the social rented sector. However, in this five years, it means that the properties buyers number will also decrease. Even, their properties prices will be fallen down to attract new tenants to rent the property developers' houses to live.

I believe the property purchase demand won't be raised, due to the rent subsidiaries assistance benefit to this country' people in this five years period. Hence, many people will choose to rent houses to live in this country. Even, their salaries are increased. It is not easy to influence their rent choices. Hence, the properties sold number must be decreased.

In fact, this country's properties or providers must not earn more profit from properties selling in this five years. Hence, long term rent subsidiaries assistance housing benefit to the country's people, it can help the low income working age group people who can have enough money to rant the better quality houses to live to improve their housing and living qualities. But, it also influence the middle or high income working age group people . When they have effort to buy houses, the rent subsidiaries assistance implementation will influence their mind to choose to rent houses to live in this five years, because they can pay lesser rent amount form government's rent subsidiaries assistance. It will reduce the middle or high income working age group property buyers number to the landlords in society. Hence, rent subsidiaries assistance will decrease the society's house buyers number, but , it also increases the tenants number in the rent subsidiaries assistance period. It will bring negative impact to reduce the house buyers number and the property developers will reduce houses selling profit, due to the house tenants number rises to influence the choice or house buyers number decreases in the country's house renting or buying markets.

However, rent subsidiaries assistance can also bring these benefits to the country's society, such as it encourages greater mobility within the social rented sector, because the properties developers will increase the small, medium, large different sizes space apartment number to let them to choose to rent, e.g. one small or large bedroom, or two small or large bedrooms, or more than two small or large bedrooms of apartment to let the tenants to choose to occupy. So, the property market will increase many different occupying space size of bedrooms number to let the country's tenants to choose to rent to satisfy their living needs. Moreover, it can make better use of available social housing stock, when there are many houses to used to be rented intention only. It can improve work-incentives for working age group tenants in the society, due to many people choose to rent houses to live in the country, they won't need to pay long term installment payment to buy the house in long term installment period, e.g. they won't need to pay twenty to forty years , even more installment housing payment to the housing developers. Hence, the tenants will reduce finance burden when they choose to pay cheaper rents from their government's rent subsidiaries assistance long-term. It will influence the working age tenant group 's saving habit to be better any more, due to they will have more extra money to save to banks when they pay lesser rent from government's long term rent subsidiaries assistance per month. Because of the high proportion of tenants in respect of rent subsidiaries assistance housing benefit may have a sizable positive influence to encourage their saving habits on the saving behaviors or work-incentive behaviors both.

So, the rent subsidiaries assistance housing long-term fixed amount paid benefit will continue to provide support where apartment is suitable for the needs of the tenants, and it will provide an economic incentive for working age tenants to move out from smaller space size properties to their apartment is consider larger space size apartment than necessary. When rent subsidiaries assistance can support the working age tenants to rent larger space size apartment to live to bring both working and economic incentive to improve their living of quality needs.

As a result, working age rent subsidiaries assistance housing benefit in the social rented sector will face similar choices in the private houses rented sector, instead of public houses rented sector. Rent subsidiaries assistance can also encourage the working age tenants either to occupy appropriately sized accommodation , when they had paid cheaper rent in a period, then they expect to buy the accommodation to live or they pay towards another accommodation which is larger than the needs of their long-term rent accommodation choice.

Hence, if the working age tenant still chooses to live cheaper rent accommodation , the choice is to move and a lower rent is payable. This will help to provide an additional work incentive. Because the working age tenant does not choose to pay higher rent to rent large space size or more bedrooms or better quality of apartment to live, then he / she will have more extra money to save to bank, even, the working age tenant can earn rent subsidiaries assistance housing benefit at lower income level when he / she still chooses to pay less rent amount to live poor quality and / or small space size of lesser bedrooms number houses to live.

Hence, rent subsidiaries assistance seems that it can bring work and economic incentive both benefits to the low income level of working age tenants when they still choose to pay lesser rents to live small space size or lesser bedrooms number or poor quality of houses to live in their countries. It can encourage them to hard to work to save more money at bank to prepare to have enough money to buy another new or second house to live. Moreover, rent subsidiaries assistance will bring social attribution benefit from these young working age workers ( tenants) in any countries' societies.

In conclusion, rent subsidiaries assistance implement can bring benefits to the low income working age tenants, but it can also bring low housing selling profit to the country's property developers. Hence, it depends on the country's policy implement whether it is beneficial to economic growth or improving low income working age living beneficial aims. If it felt that economic growth is more important to compare low income young working age living improving benefit in the year. The, rent subsidiaries assistance implementation is not right time to implement in this year. Otherwise, if it felt that low income young working age living improving benefit is more important, then rent subsidiaries assistance policy is right time to implement in this year. Hence, every country's government need to consider when is the right time to implement rent subsidiaries assistance policy in order to keep it's country house market's renting and house buying supply and demand number more balance between the house buyers or tenants and the house developers. It is beneficial to the country's long term economic and improving of living quality development.

2.2artificial intelligence estimate how much rent is the most attractive level to influence you age house renters to choose rent in preferable

Can artificial intelligence estimate how much rent is the most attractive level to influence you age house renters to choose rent in preferable? Can rent subdisiary assistanc attract householders to choose to rent appartments to live easily? Some country governments indicate that rates of residentil mobility tend to be lower among homeowners than renters. It is possibly reflecting the higher moving costs to owner, occupied housing, which my cause the the old house owners won't change their old houses to move to new houses to live easily, if the higher moving costs are charged to them higher costs . So, it seems that it is not easy to attract the homeowners to choose to pay rent to live when they are borrowing mortage loan to pay installment to live the houses, even the country government has rent subsidiary assistance to persuade or attract them to sell their houses to choose to py cheaper rent expenditure from rent subsidiary assistance support. When, the country has many homeowners who are not easily to move to another new appartment to choose rent payment, even the country government can provide rent subsidiary assistance to them.

So, it can explain that the country has the increase in homeownership rates, it mens the tendency for homeownership rates rise with young age, it implies that th country's unemployment rate is reducing, due to there has young population ageing group who is working in society. So, it can also explain that why the homeowners who borrowing house mortage loans (debts) to pay installment to live the houses, they won't easy to choose to change to pay rent to live. Although, they have government rent subsidiary assistance to support them to pay the lower housing market rent for long term.

It brings this question: How to attract the young working age population homeowners to choose to pay rent to live from their countries‘ government rent subsidiary assistance, when they are borrowing housing mortgage loan to pay installment to live their houses?

A significant proportion of the changes in aggregate homeownership rates is. However, not explained by changes in the characteristic of the population , a possible chnge in the relative attractiveness of owner-occupied housing has possible influenced to attract the young working age homeworkers to choose to borrow housing mortage loans to pay installment to buy the houses to live for long times, e.g. ten to twenty years installment time, even more than twenty years installment time. It is possible that the appartment location has many transportation tools to provide them to choose any kinds of transportation tools to go to work conveniently, or the appartment location has many shopping centers or entertainment facilities, e.g. public hospitals , public swimming pools, public libraries, cinemas, public gardens, restaurants etc. facilities to attract them to choose to live the location. So, even the country government can provide rent subsidiary assistance to support them to rent high quality of provide houses to live. This rent subsidiary assistance policy can not influence the young working age population homeowners to change their mortgage loan installment houses to move to another better quality of houses to pay lower market rent to live, even their government has rent subsidiary assistance to persuade them to live another housing location. So, the attractive old housing location and natural environment factor and low young age unemployment factor both will influence the young age homeowners ( borrowing mortgage loan installment homeowners ) still choose to live their old house locations. Even, their government implement rent subsidiary assistance to attract them to pay low rent market valuation to rent another higher quality of houses to live.

Hence, if the country government can provide the rent subsidiary assistance house locations which can be close to the shopping centers, transportation tools, public facilities, restaurants etc. service facilities to satisfy the working young age population homeowners ( borrowing mortage loan installment homeowners). Then, it is possible to attract them to choose these renting subsidiary assistance houses to live as well as the rent subsidiary assistance houses locations can be close to the locations (places) which have more offices, factories etc. working buildings to let them to walk to work conveniently. Then, the unemploymwnt young age homeowners can also be influenced to change their present borrowing mortage loan appartments to move out to rent these houses which are close to these working places more easily. The reason is due to they need to spend less time to go to work every time for the unemployment young age homeowners, if the country government can provide the rent subsidiary assistance houses which are built to close to th working places. Then, these renting house locations are close to the working places, it can influence them to choose to rent these houses to live more easily if their present borrowing mortage loan installment of house locations are far away from any working places.

The final factor to attract working young age homeowners to move out to rent another rent subsidiary assistance appartments to live more easily. It is the rental income maximisation strategy. What is the income maximisation strategy benefits to working young age homeowners ( borrowing mortage loan installment payment homeowners)? Why does this strategy can attract them to choose to rent?

This strategy benefits include as below:

The working young age homeowner is intended to be revenue self financing, when he / she is chosen to rent another new appartment by government's rent subsidiary assistance. Consequently, combination of significant additional rental income and borrowing will be required for stock investment and to finance other activity. The rental income to various sources and it can be adversely impacted by non-payment of charges. So, low market rent payments or rent subsidiary assistances are supplemented by recovered from current tenants ( the working young age population), and subsidiary assistance planning can let the working young age population to pay lower market rent payment to the country's government for either ( public houses ) or lanlords ( private houses) in arrears rent payment later. Hence, they do not need to pay the lower market rent of this rent subsidiary assistance support immediately.

They can pay lower market rent in arrear to maximize longer arrear rent payment due to the maximisation three years period. So, the working young age tenants can save income, when they pay lower market rent ( rent subsidiary assistance) in whole lating arrears whole amout is up to three years rent payment strategy. When they choose to leave their borrowing mortage lown installment payment present houses and move out to choose pay cheaper rent to rent another new renting appartment to live from their country' government rent subsidiary assistance support.

Hence, this three year arrear lower rent market subsidiary assistance strategy can reduce the working young age tenants significant financial pressures, due to the impact of welfare reform measures ( allowing them to pay lower market rent maximisation in arrear up to three years period later) by this rent subidiary assistance strategy. Consequently, I believe that this rent subsidiary assistance maximisation three year rent in arrear payment strategy can attract them to choose to move out to choose to rent another new appartments to live more easily.

2.3 artificial intelligence analyze rent calculation method can increase renters number

Can artificial intelligence analyze rent calculation method can increase renters number ? If the country government implemented rent subsidiary assistance to attract its resident to pay lower rent market valuation to attract them to live. For long time, when they will live the renting appartment for one year, even more time, whether the renters will change their behaviours to feel the renting appartmentd are similar to homeowning apprtment to live in order to decide to live th renting houses long time.

In fact, rent subsidiary assistance implementation has benefits to solve the housing raising prices challenge, such as Hong Kong has many people re living in this small city. But, the housing prices are continue rising. Hong Kong property developers aim to earn more profits when they can sell many houses. Although, these property developers feel Hong Kong has many people who will need to buy houses to live. When Hong Kong has many people demand houses to live. But, the private houses are limited to build to supply to satisfy their living needs.

So, the Hong Kong property developers will raise housing sale prices or housing rent prices in order to sell or rent to them to earn more profits. As this Hong Kong housing case, if Hong Kong government can implement rent subsidiary assistance policy to let many Hong Kong resident to pay lower market rent valuation to rent any private houses or public houses to live. Then, it can possible encourage many Hong Kong people to choos to rent appartments to live in Hong Kong. The housing welfare to Hong Kong people which will cause many Hong Kong resident do not choose to buy houses to live in Hong Kong. They will be possible to choose to pay lower rent subsidiary expenditure to rent to rent either low quality of public houses ( HK government owning) or high quality of private houses ( private property developers owning) to live in Hong Kong in possible.

If rent subsidiary assistance can change Hong Kong resident to choose to rent houses to live, it can also bring renting housing shortage challenge in Hong Kong. Because Hong Kong resident only hope to pay long term rent to rent any appartment to live in Hong Kong. When the renter feels the appartment is old or another reasons cause the renter expects to find another new appartment to rent to live. Due to Hong Kong government can provide rent subsidiay assistance, then it is possible to encourage the Hong Kong renters only choose to rent any appartment to live in Hong Kong long time. Hence, many Hong Kong resident won't need to borrow mortage loan to pay installment to buy any houses to live, if they choose to pay long term low market rent valuation to the landlords ( private house owners) or the Hong Kong government ( public house owner) to live in Hong Kong.

It brings this question: Can rent subsidiary assistance change renters behave like homeowners?

I shall give evidences to explain as below:

Many countries, e.g. UK, US developed countries have restrictions expanded to majority, center central cities, where high house prices threatn the well being of not only most residents, but the nation's economy as a whole. So, all US, UK have restrictions housing sale prices limited to manage th property sellers‘ house sale prices to avoid householders feel burden when they choose to buy any houses or mortgage loans to pay installments to any houses in US or UK both property markets.

Why do the US, UK housing become to hard to build in these traditionally development, friendly cities, Such as New York, Washington, London big cities? The reason is because that despite supporting houses supply citywide in UK, US residents individually have an incentive to defect and block new housing proposed for their own neighborhood. So, it implies many UK, US residents do not hope to live their houses which are close or near to neighborhood.

As this reason, it explains that why many Uk, US residents do not expect to live in big cities, because big cities have built many tall appartment and every appartment is built to close or near to another or other appartment. Otherwise, US, UK country side, there are many one to three floors independent self built wood houses and private car parks at homes which are built on every independent private land as well as the distance between every independent seld build wood house which is so far awary from each others on the countryside roads. So, UK, US city side houses are very attractive to UK, US residents to choose to live. Such as this reason, it explains why it can encourage UK, US residents choose to pay long term low rent market valuation to rent these kind self built or send hand wood made houses to live in UK, US countryside if UK, UK can provide rent subsidiary assistance to support them to live long time.

When, these both US, UK countryside residents who decide to pay long term low rent market valuation to live these countriesside wood made houses. Also , when they are living one year at least in here. They will possible feel that they are UK, US the country side independent wood made houses homeowners more than renter role. So, these UK, US countryside independent renting house lands and renting private car parks and renting private wood houses which all will let them to feel that they are these renting wood houses; homeowners. So, they will not feel that they are renters in these both countries' countryside independent housesholders. Moreover, when these UK, US countryside wood house renters choose to pay rent more than ten years or twenty years low rent subsidiary assistance market valuation to rent these countryside independent wood houses to live. It is possible to influence them living behaviors are like to homeowners, and are not like to renters.

Hence it also explain why HongKong residents can pay low rent subsidiary assistance to rent low rent market valuation of either high quality of private houses or low quality of public houses to live long time. Then, their living behaviors will be influenced to do any similir homeowners' behaviors at homes, because Hong Kong long time housing renters won't feel that they are renters role to their houses, they will feel that they are homeowners to their houses in Hong Kong city.

Hence, rent subsidiary assistance policy can help any countries residents to feel that they are like homeowners more than renters when they are renting the houses to live long time. It also explains why they do not need to move out to another new house to rent or buy to live, due to the renters feel that they are the houseowners actually. So, rent subsidiary assistance policy can lead ot influence householders' minds to choose to pay long term rents to rent high of low quality of houses to live in the country. Then, it will influence the country's houses prices can not be increased more easily , because the property buyer number will be influenced to reduce in the country. Due to the country has many residents choose to rent long term houses to live in the country.

When do renters behave like homeowners? I believe that renters will feel their behaviors like homeowners when themselves country can provide rent subsidiary assistance to support them to pay lower rent market valuation to choose to rent public houses or private houses to live in long term period. For US housing prices case example, its houses prices are experienced to change dramatically. In New York and Los Angeles cities when nearly tripling in San Francisco. Driving this appreciation is an inability of new housing supply to keep up with housing rent or purchase demand, causing the price of existing units to increase. So, if US government can provide rent subsidiary assistance to let residents to pay lower rent market valuation to rent any US public housing ( poor quality) or private housing ( better quality ) houses to live in US these cities for long term. Then, the low wage of US workers are no longer migrating to high wage cities, due to they have afford to pay lower rent market valuation rent to rent any public houses ( poor quality ) or private houses ( high quality) houses to live in low wage cities long tim from US goverment rent subsidiary assistance support.

Consequently, US government will reduce many societal and housing problem, when many low wage workers have afford to pay rent to live either high quality of houses ( private houses) or low quality of houses ( public houses) in any low wage cities in US for long time. So, they do not need to forgive their jobs to migrate to high wage cities to work in order to raise their wage effort to pay high rent to rent houses in low wage cities in US more long, because they won't need to rent any houses to migrate to high wage cities to live in US. Otherwise, if US government can implement rent subsidiary assistance to share these low wage workers' rent burden to help them to rise effort to rent any public houses or private houses to live long time in low wage cities. Then , these low wage workers won't need to migrate to high wage cities to find new jobs to work. They can have enough effort to pay long time rent, even these low wage cities ' employers can only provide lower wage jobs to them to work.

Hence, rent subsidiary assistance can solve low wage workers' social competition challenge in US low wage and high wage cities' both job markets in US, because they do not need to migrate to high wage cities to find jobs to work and they need to pay higher rent to rent houses to live in high wage cities in US. Moreover, they will encounter difficulties to seek jobs in high wage cities because there are many high wage similiar skillful workers who will also need to find any kinds of similiar skillful jobs to compete to the low wage migrating workers from low wage cities. So, it will raise the competition to high wage cities workers when the low wage cities workers choose to migrate to the high wage cities to find jobs to work as well as they also need to live in high wage cities to pay higher rent when they need to live here in US.

Hence, US government can help the low wage workers to avoid to migrate to high wage cities to find jobs and live, if it can provide long term rent subsidiary assistance to support them to pay lower rent market valuation to rent either public or private houses to live in lower wage cities in US. Hence, it is one good housing welfare to US low wage residents or workers if US government can implement rent subsidiary assistance to help the lo wag workers do not ned to migrate to higher wage cities to find any jobs to work and live in US. Then, they can feel living security to live in low wage cities in US long time, when they are assisted to pay the rent subsidiary expenditure to let them to pay lower market rent valuation to live in these low wage cities in US long time. So , although, these low wage cities employers pay lower wage to compare the other higher wage cities in US. It won't influence them have no enough income to pay rent to live in these low wage cities in US. Due to US government can provide rent subsidiary assistance to reduce their rent burden to live in these low wage cities . So, the rent subsidiary assistance policy can encourage the US low wage workers to choose to pay long term rent to live any public or private houses in any one of US low wage cities long time. They will be influenced to feel that their behaviors like homeowners when they are living in any one of low wage cities long time. Although , in fact, they are payin rents to live these any one of low wage cities in US long time.

Finally, regional scientists, land economists and transportation planners have confirmed that a householder's location decision depends to a large degree , on access to opportunities sities. Subject to budget, time and otehr constraints, it is a common assumption that householders maximize their utility by locating in a desirable a home or long term renting appartments as need to necessary and desired activities. Such as work, shopping, and recreation as possible. Also, they indicate that the utility maximization is profoundly complex, dependent on far more environmental attributes than those that can be observed and quantified, for example, people often care about neighborhood qualities, public services, proximity to relatives and natural environment offer factor.

Hence, when the country government expected to implement its rent subsidary assistance policy to attract residents to choose to pay low rent market valuation to rent public houses long time successfully. They need to consider whether the location can have above any one of these services provision in order to satisfy the future public housing renters' living needs daily. The country government's rent subsidiary assistance policy whether which can attract many residents to rent houses or not. It depends on whether the location is chosen to build public or private houses where it can provide above services to satisfy future renters' living needs , if it expected many residents can choose to pay low market rent valuation to live the location long time and let these renters to feel that they are like homeowners more than renters when they are paying low market rent valuation to rent public or private houses at the location long time in order to threaten the property developers can not raise their houses prices to sell easily. So, location factor is an important factor to influence the country residents choose to pay lower market rent valuation from rent subsidiary assistance.

Consequently, rent subsidiary assistance policy can influence some countries or cities ,such as Hong Kong, US, UK, etc. their residents feel their behaviors are like to homeowners more than renters, although, in fact, they are paying rents to live. However, rent subsidiary assistance can attempt to explain that why this housing welfare policy can threaten the property developers' houses prices can have effort to be raised easily, due to this housing welfare policy will influence the housing renter number to be increased.

# Robotic business development life cycle stage

CHAPTER NINE

# Strategy function to organization

● Explaining what strategy means?

What does concept of corporate strategy mean ? Why does organization need corporate strategy ? The reasons may include : reducing cost, making reasonable or the most beneficial decisions or actions, earning above average returns etc. strategy may be a set of key decisions made to meet objectives. A strategy of a business organization is a compenhensive matter plan stating how the organization will achieve its mission and objectives.

A successful strategy may have these four perspectives, a plan, how do I get these; a pattern , in consistent actions over time; a position, it reflect markets, a ploy is a maneuver instead to outwit a competitor, a perspective is a vision, direction, a view of what the company or organization is to become. For minimizes or competitive disadvantage strategy example, company realizes merging with companies advantage . Although, it may not make its market leader , but it may venture into retailing will help it increase profit.

Strategy also may provide a clear understanding of purpose, objectives and standards performance to employees at all level in all functional areas. Usually, every firm competing in an industry as a strategy, because strategy refers how a given objectives will be achieved. For example, computer industry uses a differentation competitive emphasizes innovative product with creative design. For example, coeporate strategy, Coco Cola Inc. has followed the growth strategy by acquition. It has acquired local bottling units to become as the market leader. For function strategy example, pocter and Gamber spends huge amounts on advertising to create customer demand . They aim to maximize resource productivity. It is concerned with developing a distinctive competence to provide the firm with a competitive advantage. Thus, strategy may have different functions. It depends on whether the organization needs what strategy to achieve its objectives or aims.

- Why does organization need strategist

However, any organization needs one or more strategist (s) . strengths are individuals or groups who are primarily involved in the formulation, implementation , and evaluation of strategy. In a limited sense, all managers are strategists. Strategists may include: Consulants, entreprensurs, boards of directors, chief executive officer, senior management, corporate planning staff, strategic business unit level executives, middle level managers, executive assistant titles in any organizations.

Organizations needs outsourced consultants service because many organizations do not have a corporate planning department, owing to small size . Thus, outsource consultancy firm can provide this kind service to them.

Entrepreneurs are promoters who conceive idea of starting a business for getting maximum returns on investment. They are awaiting for an environment change and for an opportunity in the best interest, for example, a biotechnology firm's managing director needs to implement policy formulation in research and development department.

Board of directors are professionals elected may by the shareholders of the company as per rules and regulation of the company act. They are responsible for the general administration of the organization. They are supposed to guide the top management .

CEO is the top man, next to the directors of the board the occupies the most sensitive post, being held responsible for all aspects to strategic management right from formulation to evaluation of strategy.

Senior management from the chief executive to the level of functional or profit centre heads. They are involved in various aspect of strategic management .

Strateic business unit is diivided into different independent units and allowed to form own respective strategies. Middle level managers are operational planners, for departmental plans, as implementers of the decisions as well as executive assistant is a person who assists the chief executive in the performance of his duties in various ways, e.g. data collection and analysis, suggesting alternative, where decisions are required. preparing briefs of various proposals, projects and reports, helping in public relation. All of above positons may be any organizations‘ role in strategic plan.

● Why do SWOT ( strengths, weaknesses, opportunitites and threats) analysis can remain a major strategic tool to any organizations?

It is one straight formed methodology for making a structured analysis of strengths and weaknesses into core competences and core problems by using the core-competence tree and the current reality true. The core competences and core problems are then linked into a plan of action aimed at preserving the organization's core competence. It supposes that any organizations ought have internal strengths and weaknesses both. So, if the organization has strengths , it also ought have weaknesses. Any organization, itself ought have ability to control or avoid or threaten its any weaknesses cause as well as finds any method to arise strengths to bring itself competitive ability. Otherwise, due to external environment factor, it can not control. So, it supposes any organization can not control any opportunities ot threats when they will occur or encounter to influence weaknesses, eliminate all weaknesses that do not satisfy the following criteria.

The weaknesses that do not satisfy the following criteria: The weaknesses must exist over a period of time can not be a one-time phenomenon, the weakness must be expressed in undesirable terms, the weakness must be under the firm's control or influence . So, any firm hopes to eliminate weakness, it depends on how it causes significant damage to the company. For example, when one firm discovers that the project ought may be finished within five months. But, after four months, it discovered that this project can not be finished, if it hopes staffs can cooperate to finish this project before five months, it needs to find whether what its main weaknesses are influenced this project will delay, e.g. lack of innovation, lack of growth, insufficient attractive profits to excite staffs to work. Hence , SWOT is a strategic management tool, it consists of the analysis, decisions and actions, an organization undertakes in order to create competitive advantages. However, the next phases of the strategic management process is external and internal organization's strengths and weaknesses analysis, by conducting an external analysis , an organization also needs to identify the critical threats and opportunities in its competitive environment. It also needs to examine who external competitive environment influences its business develops in long term.

In fact, any organizations need to make the most reasonable strategic choice with vision, mission, objectives and the external and internal analysis of its

external environment influence. Hence, the strategic management process may include this steps:
From vision to mission to objectives to ( SWOT analysis, external analysis and internal analysis both ) to strategic choice ( the most reasonable choice) to strategy implementation to achive competitive advantage . So, any organizations must need to spend long time to gather data to analysis whether which it has actual internal strengths and weaknesses as well as what the present external envioronment brings opportunities and threats to influence its business development, if it hopes to implement effective strategic plan. So, it seems that SWOT ought be one step to any organization's strategic plan management process. Thus, managers have responsibilities to help their organizations to try to " fit" the analysis of externalities and internalities, to balance the organization's strengths and weaknesses as well as environmental opportunities and threats . For one car sale manager hopes to find methods to solve its car low sale problem. In the SWOT analysis, it may have these questions: Why does the performance of the car firms in the same motoe sale service industry, operating under the same competitive environment? Which tangible resources of the high performance motor sale service firm provide sources of competitive advantage and subsequent superior motor sale service firm performance? How do the identifical tangible resources actually create value for a motor sale service firm in the motore service industry and provide the motor sale service firm with source of sustainable competitive advantage?
Thus, assumption of questions are needed in order to help organizations to attempt to seek the main factors influence their short or long objectives can not achieve in the SWOT analysis process. Then, they can evaluate the different factors to make the reasonable analysis to decide whether which is the main factor to influence their objectives can not achieve satisfactory . Then, they can revise their errors as well as find the most reasonable or the most right solutions methods to achieve their objectives more successfully. Because some factors influence the business actions its objectives succussfully. They may include: poor organizational behavior, e.g. worse staff performance, working attitude, lazy , they do not enjoy or feel bore to so their work, they feel salaries are not reasonable; poor strategy, e.g. the business ought not expand more branches at this moment rapidly, the business ought chose partnership , it is more suitable to compare sole trader formation, the organization ought advertise to raise its brand awareness to let public to acknowledge etc. wrong strategy implementation. So, SWOT

stragegy role may also help any organizatons to revise whether they have errors in order to find the most reasonable factor to cause their poor performance or low profit etc. effects.

- What is strategy and strategic management to future managers in organizations?

Are they understood and recognized? However, I believe that the development of organizational strategy depends on understanding the perceptions of their managers on what strategy and strategic management actually is. The identifications of perceptions of future maangers will need have some insights , opinions and knowledge on the organization's this matter reflect the efficiency and effectiveness of the strategy related learning proces in themsleves organizations. So, I believe that, the manager needs have enough knowledge about how to manage the kind of business if he/she hopes to become the organization's proficient leader, for example, one supermarket business CEO ought own part supermarket operation experience, when he/she has practised supermarket operation experience to know how to manager teams cooperation efficiently, e.g. cashiers, food promoters, food warehouse delivers, grocery shelf putters, fresh fish and fruit pick up keepers, family daily e.g. tooth paste, bath daily products, washing cloth products shelf putters staffs. Then, the supermarket store manager ought have excellent managing ability to manage the supermarket different sale teams to cooperate efficiently. So, owing the kind business managing experience to the manager will be one main factor to influence the business to grow or expand more successfully. Thus, how management develops strategies to guide how an organization conducts its business and how it will achieve its target objectives . The manager himself/herself managing ability and the organization's target objectives can be achieved, they will have close relationship . It is management's responsibility to adjust negative conditions by undertaking strategies defense and managerial approaches that can overcome adversity. However, the essence of the good strategy-making is to build a positive strong and flexible enough to provide successful performance despite unforeseeable and unexpected external factors.

Thus, the five tasks of strategic management may include as below: First step, developing a vision and a mission. It means that any firm ask is " What is our business and what will it be ? What is our business and what will it be ? " Managers need to develop the next five to ten years a clear mission to his organization needs to achieve. A clear mission can establish the organization's future effects and outlines " Who we are, What we need to do and where we are going ? "

Next step. setting objectives, or mission statements can achieve performance targets more easily . Objectives serve as for tracking an organization's performance and progress . A desired performance can pushes an organization to be more incentive, how to improve its financial performance, and its business position. Objectives may have short, middle and long time. Objectives may have two kinds. One is financial objectives, e.g. measures as earning's growth, return on investment and cash flow. The another is strategic objectives, it provides consistent direction in strengthening a company's overall business positon. They relate more directly to a company's overall competitive situation, such as growing faster than the industry's average and making gains in market share.

Then, crafting a strategy step, it is a SWOT analysis to find how to achieve organizational mission. Thus, strategy will be the most important part in order to let the organization to achieve its short, middle and long objectives more easily. For motor industry example, global competitor is competitive environment to motor manufacture industry, if the motor firm can not innovate its any kinds of motors, then it can not attract car buyers to choose its cars to drive. So, in vehicle manufacturing strategy aspect, if the vehicle firm hopes to raise its any kinds of car sale number. How to innovate to manufacturer " new design cars" which may be one important factor to influence any one motor firm sale growth in success. Then, car industry strategy ought concentrate on how to improve the car design to be more attractive. It is " the growth of motor design techniques concep strategy to any nowadays car manufacturers ". It seems that is SWOT analysis to car industry internal strengths and weaknesses analysis have more influential to compare external environment opportunities and threats to any one car manufacturer seller ought not only consider how to train car salespeople sale skill. They ought consider how to provide the useful opinions from car customers' design feeling to car manufacturers, in order to assist them to attempt to design any the most attractive car design to satisfy car buyers' needs in this global car competitive market. So, car design will be one important factor to influence car sale growth. Strategy ought focus on " how to innovate car design" to satisfy car buyers' car design pursue.

On conclusion, any kinds of businesses must have themselves characteristics or features. So , management ought need to consider how themselves business features or characteristics to decide the most reasonable ot the most right strategy implementation in order to achieve their objectives or missions more easily. So, lacking any strategic organizations ought be difficulty to achieve their missions or objectives to compare owning any strategic organizations in nowadays business environment.

● What does business development strategy ?

An effecting business development strategy ought have these five steps: The first step is market analysis. Who are your clients , knowledge of your market? Second step is how to adopt for each penetration, your business needs to learn how to adopt for each group of clients, your first need to review your own capacbility. It is important that you are realistic and honest with yourselves over where clients truly sit, learn how to classify your clients into similar groups relative is the scale of the opportunity. Third step learns how to review your performance , market matrix to plot your results to help you determine your market penerstion. In addition, it will help you then discuss and consider various strategies for growth. By potting your clients you will get a sense of where your strengths and weaknesses are against the opportunity that total market offer.Fourth step learns how to consider alternative growth strategies on the market matrix. The final step , you need to consider these questions in order to decide whic is the most effective strategy for your business. For example, which model is the most ( least effective? Why? which model work best for line managers, HR are finance, why? How might we most effectively progress from one model to the most reasonable questions? ) Then, you will need to decide how to launch new services, new products, opening new markets, how accessing new geographic territories.

● What is business model?

It is logic and provides date and other service evidences that demonstrates how a business creates and delivers values to customer. It also help how to predict revenues, costs, and profit with the business enterprise delivering that value. How does on build a competitive advantages and a super normal profit ? How the enterprise creates and delivers value to customer, and receive payments to profit easily .

In essence, a business model is a conceptual, rather than financial model of a business. An effective business model may help you to decide how to create value for customers, receive payment to profit more easily. For example, driving factors include knowledge economy, the growth of the internet and ecommerce, the outsourcing and offshoring of many busness activities, and restructuring of the financial services industry, i.e. the enterprise simply need to learn packed its technology and intellectural property into a product which it sold, either as a discrect item or as a bundled package.

The existence of electronic computers that allow low cost financial statement modeling has facilitates of assumptions about future revenues and costs. Also, the concept of a business model has no established theoetical in economics or in business studies. Economic theory assumes that trades take place around tangible products : intangibles are the best. For example, inventions are often assumed to create value naturalty and enjoying protection of patients, firms can capture value by selling patients to market, i.e. the publisher sells the another's books, the books can help it to earn high level of royalty income. In economic theory explains the publishers can create intengible value, e.g. royalty as well as tangible value, e.g. selling books.

However, business mides are necessary features of market economies, it is consumer choice, transaction cost, amongst consumers and producers and competition. It meets invention and consumer wants of new product value need and the opportunity to satisfy their needs. So, good designs are likely to be highly siutational, and the design process is likely to involve processes. New business models can both faciliate and represent innovation. For example, in te sport apparel business, sponsorship is a key component of today's business models, Nike, Reebook, Adidas and other sponsor football and rugby clubs and teams as well as royalties from sale sport related products, e.g. sport shoe, spot cloth. Moreover, business models must be over times as changing markets , technologies and legal structures to adopt the kind of business market change.

A business model achieve the logic the useful and reasonable data and evidence thst support a value decision for the customer. In practice, successful business models very often become to some degree, "shared" by multiple competitors in possible . Strategy analysis is this an essential step in designing a competitive business model, i.e. low cost strategy for newspaper advertising ( including classifieds) helps cost of generative content is easy to replicate and of many different geographically separate newspaper market in the world. So, when a country's newspaper publisher may have a differentiated and hard to initate low cost advertising strategy, but it can achieve the same time effective and efficient. Its business model will be successful in newspaper publishing industry. Hence, it seems that

business model will be any businesses' essential part in their growth strategies. If the business has none a successful business model, it will not have successful growth strategy consequently.
Growth strategy is different to business development strategy, why ? Growth strategy will mark afresh start, by having all economic actors in the private sector activity and dynamically undertake efforts to promote growth with a determination to take on challenges, when the business feels that it is the right time to grow its business. Otherwise, business development strategy is not the businessmen's feeling whether it is the right time to develop its business . It is essential part to any business expects to start.

- What does business climate development strategy ?

What are the different between business growth and business development and business climate development strategy ? In fact, business growth strategy refers any businesses start up in beginning from earlier stage to nature stage of life cycle. Otherwise , business development strategy must not start up from beginning. It is common on the middle stage, the business hopes to develop its market share or new market to be more. So, it needs to find whether what its SWOT in order to develop its new niche market more successfully.
Hence, it brings this question? Business climate development strategy is on the beginning or middle or mature stage in life cycle? I shall explain as below:
A business climate development strategy means that it is one targeted policy tools appear to favour medium-sized , well established industrial enterprises over younger, small enterprises with high-growth potential operating in the services. Hence, any governments may attempt to follow the business environment to implement any methods to help small size businesses to grow up easily in the beginning stage. In general, implement to business elimate development strategy challenges may include: Lacking of coordination means that there is a disconnect between business and innovation support policies on one hand of investment localization on the other. However, to solve this challenge, governments may encourage these organizations to participate , such as technical centres, laboratories and training facilities can act as catalysts for industry, sector aggregation and support the establishment and specialized investment zones. Hence, the difference between business climate development strategy and the other both strategies. Business climate development strategy is any countries' governments attempt to follow the business environment at the moment to

find the most effective methods to help any small size businesses to grow up more easily in the beginning stage.

How to implement business climate development strategy more successful? The key recommendations may include: To emergy from the workshop was further strengthened in an effort to close the policy gap and create synergies between programmes. Building on the experience of regional investment centres, the context points would identify an small middle size's (SMS) organization, and help enterpreneurs to establish a network of SMS bisiness centres across the country. So, they could not as single -window contact points, storing and channelling information abour all the government's SMS business programmes. However, a successful business climate development strategy should build on the analyzed risks and rewards of informal business operations and aim at modifying the behavior of economic agents ( enterprises, employees and customers) through a combination of incentives and penalties.

Any governments may attempt to develop a number of measures to promote and suport innovation and upgrade technology in the private enterprise sector. In order to enhance business climate development strategy to implement in success. They may include: establish a system of communication and cooperation between the institutions and private sector organizations operating in the area of technological upgrading, innovation, financial, technical standards, public education and training . It would be useful to conduct an evaluation of the enterprise Europe network's impact in order to learn from the lessons of the country's more successful sector-specific centres. It is the national strategy for innovation, technology upgrading and investment encouragement. Finally, a critical element in an effective innovation strategy is the establishment of links between support services and programmes and access to funding to support any business founder hopes to develop himself business in success in order to adopt business environment climate change more easily.

However, in general, small or medium size business will encounter these challenges when they hope to adopt the business environment climate change to grow their businesses more easily. Their challenges may include: The lack of economies of scale, which limits their ability to invest in fixed capital and technological development, proportionally higher costs,, which increases the impact of the legislative and regulatory framework, information lacking which limit access to external financing, limited resources for internal training and human capital developments. In general, any governments need to help the new business to solve these challenges in order to develop business climate development strategy more easily. They may include: innovation technology centres and networks as well as financial support for innovative SMES. Hence, business climate connection with larger foreign enterprises through active government support, the policy objective is to enhance SME access to international markets, skills development, finance and technology to any businesses‘ beginning stage in order they can grow up their businesses to nature stage in their business life cycle successfully.

Hence, whether which firms hace real need to get government's business climate development strategy assistance. I believe that the government needs to assume the firm has these challenges in order to ensure that the firm can achieve the requirement to get this business climate development strategy assistance. They are needed to assume on basic these factors: A company must grow and pass through all stages of development or die in attempt, second the models fail to capture the important early stages in a company's origin and growth. Third, instead of annual sales , although some mention number of employees whether it is more or less factor, government can not ignore other factors to decide whether the firm can be accepted to implement climate business development strategy assistance, such as value added, number of locations, complexity of product line and ratio of change in products or production technology etc. factors to decide whether the firm is suitable to be accepted to get business climate development strategy assistance from the government in order to avoid waste time and resource and money to desing any kind of business climate development strategy to assist the firm to develop in the beginning.

● How to implement successful organizational downsizing strategy?

What are the effects of downsizing on organizational performance? What is the most right time to downsizing to the organization ? When one organization has grown to the bigger size , e.g. more revenues, this year than last, a larger workforce, greatest market share, downsizing strategy ? If an organization did not grow, it was viewed as stagnating and upproductive in the non-growth life cycles stage, it implements downsizing strategy to influence its performance to be worse.

Organizational downsizing strategy is one part of the management of an organization and designed to improve organizational efficiency, productivity and/or competitiveness. Downsizing means to reduce organizational size, e.g. staffs number reduces expenditure reduces, cost reduces . Downsizing is an intentional set of activities, it differentiates from loss of market share, loss of revenues or unwritting loss of human resources organizational decline. Also, downsizing usually reductions in personel, such as transfers outplacement, retirement incentives reduction, byout packages, layoffers. This reductions in personnel may occur in one part of an organization, but not in other parts, e.g. in the production function, or not in the engineering function. Finally, downsizing may effect work processes, e.g. fewer employees are left to do the same amount of work, and this has an impact on what work gets done and how it gets done. However, instead of downsizing of reduction employees number aspect, it may also occur on other accepts, such as selling off, transferring out, merging businesses or altering the industry structure. It aims to improve organizational performance. Labeled workforce reduction strategies, focused mainly on eliminating headcount or reducing the number of employees in the workforce. It aims to early retirements, transfers and outplacement, by-out packages. This kind downsizing strategy whether it can bring performance improving benefit to organization or not in long term? Less employees work whether it will still improve performance, although the organization can reduce salary expenditure . Otherwise, if the organization does not reduce staffs number, it chooses to workforce reduction, work redesign and systemic strategies , whether it will be better than staffs number reduction strategy ?

What are critical success factors influence any organizations' strategic downsizing success implement ? Addresses the rationale utilized by firms to downsize, the expected outcomes in terms of economic and human consequences, and specific strategy. Also , downsizing tactics, human resources as assets to cost planning, participation, leadership, communications and support to victims . survivors are examined to any attempt implementing downsizing organizations.

In past organizations, when many blue-collor workers are also to trade off wage freezes for jobs security. White-dollars workers in the lower ranks of white-collar workers are often dismissed by downsizing, due to firms are increasingly forced to cut costs, restructure, and reduce their labor force. Instead of western countries firms are popular to accept downsizing . Downsizing has even become common in industrialized countries, such as Japan and Sweden, restructing in the 1990s led to employment reductions in industry and thus, increases in the levels of unemployment. Hence, downsizing may cause low ranks of white-collar workers feel job security lose in any time when they are working in any organizations, because white-collar workers are different to blue-collar workers have unions protection. However, there are three perspectives from when downsizing can be reviewed : The industry level, the organization level and the individual level in terms of industry or global perspective, it may include , mergers, acquisitions, joint ventures, the organizational and strategy level may include how to implement downsizing and the expected bebefits of downsizing on the firm's performance, efficiency , and at the individual psychological level, it includes employee himself/herself stress, negative emotion feeling , due to he/she is dismissed. Hence, any organizations need to considerate how downsizing brings negative emotion to influence every dismissed employees. Because , their leave which will influence the continue working employee's emotion feel fear to be dismissed in next. If the present employees often feel stress to work, then their performance and efficiency will be influenced to worse. So, any organizations can not neglect to care the current working employees individual emotin in order to avoid low efficiency and poor performance to their organizations. Becaus every employee will have possible to be unreasonable dismiss, due to downsizing organizational influence.

On possible reason for this occurrence, is that firms poorly planned or carried out earlier downsizing projects and hence must remedy past facilities. So, one planned downsizing strategy will avoid negative emotion to influence current employees' works. But, factory workers will have possible to encounter dismiss , due to technological improvements, e.g. robotics can reduce to have additional workers rather than replacing the existing employees. So, when the factory begins to apply robotics , then employees number will be reduced. It is technological manufactuer causes downsizing to factory workers reason. It is due to raise productive efficiency factor , more than reducing cost reason to cause downsizing need to any organizations.

The term downsizing was first used reforcing to strategies to reduce personnal. However, it was become more and more relevant , its scope has been expanded and noe refers to a wide range of management measures towards better adopting on organizations to its environment ( Gandolfi & Hanson, 2011).

reference

Gandolfi , F. & Hanson, M. (2011). Causes and consequences of downsizing : towards an integrative framwork, Journal of management & organization , 17(4), 498-521.

In general, it is needed to implement strategies, due to the organization feels that without achieving the required organizational changes, this failing toobtain the desired results ( Magan & Cespeses, 2012).

The downsizing methods may include: retrenchment specialized production, concentrating activities until economies of scale have been achieved. Downscaling strategy is toward again reducing in a smaller differentiation of activities in the value chain , it aims to keep the organization to reduce complexity in the organization. For some organizations had begun to implement robotic factory, because they expect that manufacturing robotics can help they to specialize production, raise productive efficiency. Hence, they only concentrate on keeping the proficient workers, they can cooperate to robotics to work in order to raise more products number efficietntly every day. So, the low skillful workers will be dismissed and they will re-employ the owning control manuacturing robotics skillful new workers to replace them. So, future manufacturing robotic manufacture plants causes downsize, it will be one good example for specialized production, concentating activities until economies of sale reasons to cause downsizing factory workers need to the owning manufacture robotics plant organizations. So, downsizing has an impact on raising productivities on specialized production and concentrating activities until economies of scale aim more than reducing cost to the owning manufacturing robotic factories organizations.

Thus, downsizing activities aim to improve organizational efficiency, productivity and/or competitiveness that affect the size of the firm's workforce, costs and the work processes. Downsizing may include: building-down , de-hiring, de-recruitment, reduction in force, re-sizing and right -sizing. So, in macroeconomic factors view, global competiton , technological innovations ,a change in business strategy retains competitive advantages may cause why some organizations decide to implement downsizing strategy.

However, one successful downsizing strategy implements to any organizations, organizations can not only consider themselves benefits, they also need to consider the psychological contract between employer and employee as new mutual expectations on workplace environment. Frequently described as re-organization, restructing, downsizing or real sizing, the human resource effects of these changes have often been very destructive to individual lives, employment relationships and organizational efficiency. If mployees recognized that their company was creative and consistent action to pressure their employment ( security, trust could be reestablished and the success of the adoptive strategies. They can feel that their organizations decide to achieve the downsizing strategy is very reasonable more than unreasonable strategy in the right time. For example, when the organization decides to implement downsizing srategy before, they may enquire to their staffs opinions and acknowledge what their emotions, e.g. information on when change provided, staff views on the change are sought and are acted upon, staff have the opportunity to voice disagreement, support from manager during the change, to let they feel that change process seen as fair and equitable to let staff feel job security during the change process, and staff are trained to meet new job roles. So, all these factors may influence present employees have confidence to continue to work in your organizations after downsizing strategy is implemented. So, any organizations can not neglect to consider their present employees' feeling or emotion in order to avoid many staffs decide to leave their organizations after downsizing strategy in implemenation later.

reference

Magan , A. & Cespeses, J. (2012). Why are Spanish companies implementing downsizing. Review of business 32(2), 5-22.

● Business growth strategy

What factors cam affect the performance and growth to small businesses? Why and how obstacles are problematic for growth? How these differ between micro, small and medium sized businesses? How the obstacles are shaped?

Any businessmen had a substative growth ambition, but it can not represent that growth ambitin must sicceed to grow up their businesses. However, they must need to solve challenges when they expect grow their businesses successfully. The unpredicted external environment , include market changing and the vision of the owner and their attitudes towards growth will influence whether their businesses can grow up in success.

Hence, if the new business can keep negative growth, it ought may suceed. Some strategies , business owners need to consider that these factors will obstacle to their growth, such as during a recession, their businesses ought be difficult to growth, strategic planning is only useful when the business has a definit objective in mind, investment in research and development is too risky, expensive and difficult for a small business, there is no way,we can improve cashflow situation, factoring is only useful, if you ave in trouble, employees do not want formal, pay-related incentives and they are no use in helping business grow, we do not need to engage the staff in a structured, involving way, we can not get recruits to fit our needs, our business don't need to restructure our management as our business grow.

● How whether what obstacles can affect small business growth?

I assume small businesses have general staffs number with 50 or less than 50 staffs . Also, growth ambition was higher among younger business owners. The business owner personal poor time management factor may be one obstacles, for example, a lack of management time was rated to be the most difficult obstacle for potential exporters, which is something of an obstacles for the significant exporters, little knowledge of how to export and difficulty in finding customers also attract higher ratings from potential exporters, e.g. the fear of payment problems, the cost of exporting and being too small to export are rated as being less significant obstacles by potential exporters and their perceptions are not to distant from the significant exporters. So, lack of management time and little knowledge of how to export may be the business ower's significant obstacles.

However, many small businesses are facing values number reducing challenges. How adoption to improve sale performance? Sale improvement strategy may include: appreciated new customers, more advertising, devised a new marketing strategy, dedicated sales/ marketing manager, undertaken training in marketing sales.

Overall, making this transition from being a micro business to a small business clearly requires a greater confidence in dealing with such matters, though, undertaking activity more frequently or simply the earger scale of the business necessitating increased familiarity and competence. Also, employing a professional manager can be seen as generally enabling a company to improve.

The aim of focus groups was to explore a company to improve owners' views on growth, including how they conceptialize growth, perceived barriers,the consequences of growth and personal cirsumstances and evidence of mindsets among owners which may restrict their potential business growth. In general, family owning-business strategies may include: maintain quality and higher prices, rather than lower prices raising the value added of products and shifting into a less marketplace and the emphasis towards areas where they sold direct to end clients, rather than acting as subsontractors, whose margins were being squeezed, up-selling in terms of volume or value to existing customers , e.g. a catering establishment noted then they tried to encourage customers to return, or an accountancy practice and a range of extra services.

Many owners did also acknowledge that the would likely more if they were more actively intending to grow business or if they saw evidence of potentially opportunities. Several those lacking a current plan were aware of a growing need to develop both a more strategies outlook and more formal systems, because of a general neglect of strategic thought, with a number noting that years of unplanned growth has left them realizing that the development of the busines, needed to catch up with the situation that they had found themselves in.

- What can impact on growth strategies on business?

Overall, making this transition from being a micro business to a small business clearly requires a greater confidence in dealing with such matters, through undertaking activity more frequently or simply the earger scale of the business necessitating measured familiarity and competence. Also, employing a professional manager can be seens as generally enabling a company to improve.

The aim of the focus groups was to explore in depth business owners' views on growth, including how they conceptualize growth, perceived barriers, the consequences of growth and personal circumstances and evidence of mindsets among owners which may restrict their potential business growth. In general, family ownin, business strategies may include: maintain quality and higher prices, rather than lower pruces raising the value added of products and shifting into a less marketplace and the emphasis towards areas where they sold direct to end clients, rather than acting as subcontractors, whose margins were bring squeezed, up-selling in terms of volume or value to existing customers , e.g. a catering establishment noted that they tried to encourage customers to return, or an accountancy practice and a range of extra services.

Manyowners did also acknowledge that they would likely plan more if they were more actively intending to grow business, or if they saw evidence of potentially opportunites. Several of those lacking a current plan were aware of a growing need to develop both a more strategic outlook and more formal systems, because of a general neglect of strategic thought, with a number noting that years of unplanned growth had left than realiaing that the development of the business needed to catch up with the situation that they had found themselves in.

- What can impact on growth strategies on business ?

Growth is important and key on survival of any business profit venture. Formulating and implementing effective growth strategies may enhance business pforit to any dynamic organization . Developing growth strategies to attract human resources, including increase in the sales volume per annum, an increase in the production capacity, increase in employment, increase in production volume and increase in the all of material, increase energy and power, these factors may influence the business's strategy is implemented effectively in order to achieve growth aim.

Growth strategies that a business enterprise may wish to adopt include: understanding customer expectation, service, positioning, market segementation, setting measuring market standards, relationship marketing, human resource strategy and successful communication strategy these factors may influence business growth success. When one organization ensures that it can achieve growth, it may evaluate to measure its overall performance by these several aspects, they may include sales, assets base, employee retention goodwill and increase business profits that drive investment and economic development. Business growth may introduce new products and services, or adding new features to existing products. Growth could also mean expansion of an organization in order to buy new assets develop new products or service to enhance new investments in the economy. I shall refer some growth strategies as below:

Market penetration strategy focuses on expanding sales of a company's existing products or services in an existing market. It may attract new customer for the products and increase the usage or purchase rate of existing customers , it is often achieved by increasing activities through more intensive distribution and competitive pricing promotion.

Market expansion or market development means to move it into a completely new market. This strategy is about existing product which are offered in a new market when a region business wants to expand, or when new markets are opening up, or new use is found for the existing product.

Product expansion or product development strategy means introducing a new idea into a company's existing market. It offers new products to an existing market. It tried to grow by developing improved products for the present market.

Diversification means companies with sell new products or new market. It is very risky strategy . It needs to research market to determine if consumers in the new market will potentiallu like the new products. Acquisition means the purchase of one company by another company. It may be private or public.

The new growth strategy can be used as an alternative channel. It involves pursuring cutomers in different ways for instance selling a company's products or services online . Through the use of the internet a customer can access products or services of a particular company in a new (alternative). It goes beyond envisioning a long-term success. It has to be follow these steps: establishing a value for the company, identifying an ideal customer , who is loyal to the company, defining a company's key indicators, verifying revenue streams for cost reduction, seeking competition in the external environment, focusing on company's strength ( internal), investing in talents ( effective human resources who are creative and innovation). So, managers must achieve on growth strategy to enable stakeholders not only to plan, but also to track organic growth in their revenue and allow effective and efficient allocation of resources toward a more centered effort to adapt to frequent changes in the company and the industry occasioned by technology and the differences in competition.

Hence, of growth strategies are effectively formulated and implemented according to indicators and plan, it will lead to increase a profit in that organization . Growth strategies are often called the master business strategies, they provide the basic direction for strategies action. They are the coodinated and efforts indirected towards achieving long term business objectives and profit. Growth strategies have played central roles in the expansion and profit. They have enabled organizations to increase market shares, develop new markets, and develop new products and services, so business profit will continue to increase economic development.

- Why and how can organizational life cycle models influence organizational performance?

What does organizational life cycle mean? Must any organizations have life cycle? What do the influences when the organization reachs the organizational life cycle stage? Can the organization implement any useful strategy , if had ability to known whether what stage is its organizational life cycle? I shall explain as below:

In general, organizational life cycle has three stages: Birth, young, and maturity or decline. The related goals of profit, growth and survival seem to have overall goal structures of most organizations. IN general, most organizations will experience all three stages. However, not all organizations pass through all three stages. In fact, only about one-half of all new busness, organizations survive longer than one and half years. Relatively few for profit or not -for-profit organizations survive long time to experience all three stages. I assume that profits growth is one main factor to influence any organizations whether it can experience long whole three stages in their organizational life cycle.

What is the three stages chacracteristics of organizational life cycle? In birth stage, a merger or a point venture may occasionally lead to the creation of a new organization. A organization may be either a single person expands or an entrepreneur cooperates people to help promote a new idea, product or service. The motive in both cases is usually the desire for profit. In youth stage, when professional management is taken over by a family with a controlling interest, the organization's primary goal often changing from profit to growth. The new management team wants to demonstrate its competence and growth is the most obvious aim. For example, a manager of a large organization must consider how much company's return an investment in the organization's growth stage in this new growth stage. It has these characteristics: goals become less specific, less measurable, increasing emphasis on marketing, hoping for the increases sales that will justify the expansion of plant anf acquisition of new, more effecticient tools and equipment. Finally in the maturity / decline stage, as an organization matures and starts to decline, a desire to survive which will be the organization's goal in this stage. Why does organization may experience this stage, the organization can evaluate whether it is experiencing this stage, depends on these factors, e.g. when organization increases large , its technology is complex, its structure is bureaucratic, it is financially oriented, it is greatly affected by market and social forces and it is so complex when it grows up its organization, it will increases more new departments and it will employ many extra staffs and it will create many new positions. Hence, it explains why the organization can survive long time, and it can experience all three stages, it must be more succefful to compare the another organization can not experience all three stages , because it is common that when the organization can experience all three stages. It must survive long time. Also, it means that it can earn profit growth . Otherwise, when the organization can only experience birth stage or youth stage , then it can not survive long time and without profit growth in possible.

Hence, when one organization can experience all three stages. It may be one successful profit growth organization, e.g. although one sale trade earn less profit, and its organization size is small, but the jobb trader's business can survive a long time. So, he/she sole trading organization may experience birth, young, and mature or decline stages. So, organizational life cycle model can be applied to large or middle or small size organization, even sale trader, partnership organization to help them to evaluate whether their businesses are experiencing which stage in order to implement the most suitable or reasonable strategies to help them to solve present challenges more easily.

● What is the essential elements of life cycle model to assist business development ?

Although, when one organization feels it encoounters the decline stage, it means that its organization has possible that it can not continue survive. But if it has good strategy to help itself to solve present challenges. It has chance to renew or continue to develop its business functions to be better. Then, it has chance to continue survive. Usually, when the business is experiencing decline stage, it ought decide to change its business direction in order to continue survue . For example, raising its capabilities of organizational learning and innovations, creating new profitable and vision into the renew survival stage, and increasing its competitiveness in cost. However, the decline stage is characteristics by deterriorating profits and a loss of market share. The renewing firms have a rebuild their learning and innovative capabilities and shape a new profit direction for business. The contribution links interactions of development more effective business functions to provide a tool to help the experiencing decline stage organizations to learn how to implement new strategies to solve their present challenges in order to continue survive.

Hence, what is the essential elements to help the experiencing decline stage organizations to have possible to continut survive. I shall explain as below: In old economic society, manufacturing firms whose main driver to standardize production, products and busines processes. By constrast, the new economic society, we are experiencing information business, utilize information to differentiate, personalize and dispatch over networks at an rapid speed, small as e-commerce. It is obvious tht old economy's traditional shop business model is not popular to be accepted by consumers. Consumer shopping behaviors have been changed to choose online shopping to replace visiting shops shopping behavior. So, it seems that why some businesses will experience decline stage within one year. It is possible that their traditional business visiting step shopping method is not popular to be accept to themselves businesses, it is right time.

They need to design website store to let customers can choose online shopping when they visit themselves website stores from internet. So, e-commerce can influence some businesses to experience the decline stage in short time rapidly. It may be one factor to influence any organizational life cycle stge to be shorten in short time. It is one technological innovation element to shorten any organization's life cycle in short time. So, any businessmen ought not neglect technological innovation new influence their businesses development as well as they need to continue pursue their new sale method direction, e.g. payment by smartphone shopping method, electronic commerce payment transaction payment method, in order to keep high technological or payment channel to attract customers.

The another element is how to deliver customer focused to feel differentiation, to fight for survival in the global market, a company needs to implement innovation function to cope effectively with the changes in the customers‘ needs. So, it is innovation element to the experiencing decline stage organization to help it to attempt to solve challenges in order to continue survive in possible . Innovation may include service, e.g. sale service, client service, delivery service etc. as well as product innovation ,e g. design change to the cup, mobile technological improvement, car style, design, engine, chargeable battery charge etc. or individual innovation, e.g. the fundamental assumption of operate culture changes, mindset of the top manger, CEO midset change and questioned by the capability of execution of the operations function to himself organization. How to innovate organizational learning, whether and what experience or knowledge applied in the operations function can be executed for the next innovation in order to renew the experiencing decline stage organization strategy to continue survive. So, innoviation is also one element to influence organization's survival.

- What can learn from the organizational life cycle theory?

The next element is whether organization can continue keep on learning element. In fact, there are many different factors to influence whether our organizational survival. They may include organizational internal factors as well as outside factors. In general organizations can control themselves internal to be better, but they can not control outside environment, because they can not predict when environments, e.g. economic environment, customer shopping desires, when new competitors enter market. According to organizational life cycle theory, during the firm's growth from inception to high growth, to maturity firm characteristics differ and the internal resources and capabilities of the firm develop.

Organizations encounter an unpredictable business environment which is constantly pressured by the changing effects of globalization , competition and technological advancement within the context of the knowledge economy ( Thoumrungroje & Tansuhaj, 2007).

reference

Thoumrungroje, A. & Tansuhaj, P. (2007). Globalization effects and firm perferences. Jounral of international business research, 6 (2),43-58.

During the first stage, any organization will up in a new business environment with much adaptation and try to develop a niche through, learning and innovative practices. Given the success of that survival, the organization becomes aggressive in the second stage in how to managing internal resources, effectiveness and efficiency, learning workflows, and corporate structure to accommodate the increased complexity of operations, policies are needed to implement in second stage. The third stage, business efficiency is the core, and the organization keeps resolving workplace problems and defining clear objectives of what to achieve short and long term in the business. In this stage, revenue is the key pursue aim or objective. The final stage, as a nature stge, the organization tends to maintain the business stability and spend time focusing on the status of how organizational structure, management departments cooperation strategy implementation and organizational culture in order to help the organization itself and the CEO leader himself/herself how to manage her/his organization successfully . So, whether the organization can continue keep on learning , it may be one important element to influence the organization to develop in business.

- How to develop organizations in growth stage?

Theoretical development of the organization life cycle has description of distinct stages of organization growth, Little attention to the dynamics of organization growth, such as how to grow the organization? Why can the organization grow rapidly ? I believe any organizations expect that they can grow rapidly, but the organizations expect that they can grow rapidly, but the question concens: What factors influence the organization can not grow rapidly, e.g. lacking proficient staffs, lacking high technological manufacture, lacking creating mindset or innovation, lacking suitable strategy implementation, strong new competitors enter, customers taste change etc. different organizational internal and external environment etc. factors. So, any organizations expect to grow rapidly, they need to solve the challenges to threaten their grow. For example, one mobile manufacturer only concentrates on manufacturing new technological smart phones products . So, at this stge, it ought pursue a niche strategy, presenting a very narrow smart phone product time, often a single smart phone product to a single smart phone user market. The new different design and function of smart phone products venture generally undertakes major and frequent smart phone product innovation. Major investments are needed to make in smart phone products development, robotic plant and manufacture robotic equipment is needed. So, this smart phone product manufacture firm expects to grow its smart phone share market rapidly. It must often need to innovate itself manufacturing technology , e.g. manufacturing robotics as well as designers need have good creative mindset to attempt to design any kinds of smart phones and change kind functions and smart phone design in ordeer to satisfy smart phone users' needs. When , it often have different kinds of new smart phone products design. I believe that it can grow itself smart phone productive speed rapidly, when it can attract many smart phone users to choose its any kinds of smart phone products to buy to use in preference . Hence, this smart phone manufacturing firm needs have good design mindset to attract smart phone consumers' attention if it expects to reach the growth stage in short time. If it feels that it has not increase smart phone customer number significant in this time, it may believe that its busines can not reach growth stage in the moment. Hence, the life cycle theory offers expected obstacles for each stage, which can help the firms to solve the problems and help them to attempt to find any useful strategies accordinglyly.

Hence, organizational life cycle can help any organizations to revise whether what obstacles can influence or threaten the organization itself can not grow easily, due to consumers demand have become more complex, as well as arket trends are harder to predict and competition is severe than ever, in order to survive companies, shoulf need to learn how to reach growth stage rapidly in order to raise its competitive ability to reach mature stage rapidly in short time. Becuase of the organization can reach mature stage from birth to young stage in short time rapidly. Then , it will have enough ability to avoid to reach decline stage rapidly. If it expects to continue survive or it does not need to review its organizational strategy ocnsequently. Because different stage, it will have different kinds of challenges to the organization will encounter, e.g. in the birth stage , challenges may include lacking cash flow, less customer number, without building attractive or famous product brand or image, in the young stage challenges, may include strategy is not suitable to implement effectively, customer growth speed is slow, product development or promotion challenges when the organization reachs mature stage, its challenges may include clients number begins decrease or loss old clients, product sale number begins reduction from the top level, customers feel its products are not attractive and they begin to choose to buy other similar feature of products to replace its products in market. Hence, when the organization can know whether it reachs which stage in its organizational life cycle model. Then, it may attempt to find the most suitable or the most reasonable strategy to implement in order to keep its long survival to stay on the mature stage more easily.

Learning organizational life cycle stage strategies
advantages

Any organizations may experience organizational life cycle stages from birth stage to growth stage to maturity , then it may also experience decline and/or regrow stages. But this two stages, they are not all organizations must may attempt to experience. It depends on whether economic environment how changes, organizational itself SWOT strengths and weaknesses etc. unpredicted factors to influence that when the organization will experience decline life cycle stage. It means that if the organization has very poor performance, then the organization has possible to experience decline life cycle stage in short time or long time. Otherwise, if the organizationhas very good performance, it ought not experience decline life cycle stage in short time, when it can reach mature stage in its the topest level. Even, when the organization has poor performance, so it is experiencing decline stage, but if it may implement effective strategies to help itself organization to develop . Then, if its strategies are very effective , in consequence, the organization ought may experience regrowing stage to re-experience its mature life cycle stage again. So, it seems that if the organization can have very good performance. Client number can increase significant as well as profit can also growth rapidly. Then, the organization ought may experience long time in mature life cycle stage or it means that it will be difficult to reach decline life cycle stage. Unless, some sudden inpredicted economic environment, or strong competitors etc. influence its performance, then they will have chance to cause it experiences to decline life cycle stage from mature stage suddenly. Hence, all organizations must need to experience birht life cycle stage in beginning to this stage.

However, when the business founder starts to set up his/her business. He/she needs time to deal any difficulties,e.g. how to advertise his/her products to let customers have much knowledge, promote them to sell to market, how to implement strategies to solve organizational challenges. So, in birth stage, any organizations ought feel difficult to improve its whole performance or evaluate whether its future performance can improve to be better or can not improve or worse. Then, when the organization operates one period, it ought experience to growth stage, but it still depends on external factors to influence whether when it may experience growth stage, the factors may include: Whether strategies can be effective, economic environment is good or bad, customers purchase desire level is high or loe, cost expenditure is high or low etc. difficult factor.

So, before any organizatons may experience growth stage, there are many different complex factors to influence whether they can succeed to experience this stage easily. If the organization can not implement any effective strategies to solve its customers purchase emotion challenges, then its business is difficult to continue grow, also it means that the organization can not growor expand its business easily. Due to it can not continue to develop its business easily. It must not reach mature life cycle stage easily. Thus, any organizations can reach mature life cycle stage. It represents that its business has good strategies to solve any challenges in order to its products can attract customers to choose to buy or it can provide good service performance to satisfy clients needs to compare irs competitors in this market successfully.

In fact, it is not all organizations can attempt to experience the mature life cycle stage. This stage is any organization individual the topest stage. In this stage, the organization may have many clients increasing number significantly every year, its market can continue expand, profit can continue increases . All is the best to any organizations, if it can reaches this stage . All many organizations may only experience birth stage or growing stage . They reach this either birth or growth stage, then they have none good strategies to compete their clients number can not increase, but only decreases, profit reduces , even loss. They can not know how to change strategied to improve their performance or competitive effort to fight this competitors. Then, their businesses can not continue grow or expand. So, they have more chance to experience decline stage after either birth or growth stage only. They can not reach mature life cycle stage to attempt the topest level in whole business ( organizational) life cycle stage or process. Thus, it brings these questions: Why do organizations need to learn organizational life cycle stages? What advantages to bring if they can attempt to learn how to reach growth or mature life cycle stages easily? I shall explain as below:

● Why do organizations need to spend time to learn how may experience different business life cycle stages?

The business life cycle is the progression of a business in phases over time and is most commonly divided into five stages: Launch or birth, growth, maturity and decline or regrow. Each company begins its operations as a business and usually by launching new products or services. Because any organizations will encounter challenges in every stages . If they know what factos may help them to enter another new stage of business life cycle or what challenges may threaten them can not enter another new business life cycle stage easily. Because businessman need to learn and how adjust their business model to ensure profitability. That is why an awareness of what stage of the business life cycle , you are currently it can be helpful. Hence, how to maximize each stage of the business life cycle, the businessmen might still need to learn how to work in order to improve performance when the businessmen are experiencing any one life cycle stage. Moreover, each business life cycle stage comes still need to learn how to turn a profit and the first outlines of their governance and compliance and this is one big reason why most businesses fail at this stage.

So, I assume that business life cycle stage is similar to school examination, the student needs to spend time to learn in the birth learning stage, then he needs to test in the growth learning stage, next is examination in the mature learning stage, if the student fails, t is decline learning stage to the school. It may be due to the teachers can not teach students to learn easily. So, these are many students fail in tests or examinations. So, if the school teachers can improve teaching methods to let many students may earn high grades in tests or examinations. Then, the school may experience growth, even mature teaching life cycle stage in short time rapidly . Hence, teaching quality can improve or not , it will influence any school organizations ought feel to schools to learn how to improve teaching methods or strategies in order to let students can experience the maturity learning stage or it can also experience the maturity teaching stage. It means that it ought learn how to improve its teachers teaching service performance to satisfy students learning needs if it hopes to reach maturity learning and teaching life cycle stage in short time for itself school organization benefit.For example, the organization founder may ask himself/herself why he/she wants to start this business, learns how to manage exployees strategies? It is the learning needs in the third stage, such as maturity stage. Otherwise, in the first stage of the business entity birth life cycle is sometimes called the seed stage and a matter of iteraing, testing an learning , and trying again, knowing that the businessman is unlikely to have.

What advantages may bring to the organization if it can attempt to learn how to solve different challenges in different business life cycle stages ? What advantages to the organization, if it can know how to experience every business life cycle stage?

In fact, the business life cycle is the progression of a business in phases over time, and is consumer segments by advertising their comparative advantages and vale. For example, when the business is experiencing growth stage , in the growth phase, the business founder needs to spend time to learn how his company can experience rapid sales growth. This learning may assist his business to develop his business to enter next mature how stage easily , for example, he can learn how the rapid growth stage takes advantage from the proven sales model, e.g. online sale or traditional visiting shop sale model which is more suitable to his business, marketing model and operations model, e.g. how to advertise his product or promote his products can affect more audiences concern this will see the businessmen's jounrey from idea to start up, and if successful, how to keep to stay long time in the mature stage. Rememeber, when having a successful business model behind any businessmen is undoubtedly an advantage, it is not a disadvantage when the founder spends more time to learn hoe to run his business. In fact, he won't waste his time to learn how to improve his business in different business life cycle stages. So, a tactical plan will take any business strengths and reduces to avoid weakness cause to influence its development. So, knowing where you small product is in its product life cycle, it is important to continue to develop your business successfully. SO, any impacts of all life cycle stages, any businesses need to be considered comprehensively , for one new technological product firm example, its new technological product life cycle begins with the introduction or birth stage. The high technological product company must succeed at both developing new product and managing them in the face of changing tastes, competitors' technologies similar change. So, it is what it needs to learn in this stage for this new product technological firm preparing development to next growth stage.

On the conclusion, learning how to achieve in every business life cycle stage, it can bring these benefits to any organizations, such as : they can understand and redefine this role from a more, if the organization ony to learn sale frameworks what it could have picked up. It is not enough, because most organizations will only find that a majority of their total sale number which is to use solely supplier-specific data about the life cycle, but they neglect how to set targets to learn how to improve their sale to be better in the future time, it is one important factor explain why many organizations only reach the growth stage, but they can not experience to next mature stage more easily, due to they do not consider how to implement strategies in order to achieve their next targets. They feel often implment targets which will help them to know whether they need to how to do in order to improve their businesses to satisfy clients needs. As with any effort in your organization, communication plays a critical role, craft machine learning to predict and manage human for remote teams to work through the innovation lifecycle, serve them well. Any organizations need to learn how to satisfy any customer individual purchase jounrey ( called purchase experience) which the customer has with the organization, because when the organization can learn how to satisfy any client individual real need in any life cycle stage. On consequence, its clients number with have possible to influence increase. Thus, any organizations can bot neglect to learn how to satisfy client individual real purchase experience need in any life cycle stages because improvement to salepeople sale performance, they need spend time to learn in every time sale experience . When the organization can build excellent sale teams, then they may help it to build famous loyalty and good client relationship in order to expand its business more easily. Hence, in any businesses' life cycle stages, they must need to spend time to learn how to improve product quality service performance to bring customers' satisfactory emotion in order to expand their business developmenr more easily. So, i recommend that all small organizations expand to large size, they must need time to learn and attempt to find the best methods to solve any difficulties when they are facing in any one business cycle stage, if they want to expand their businesses successfully.

● The relationship between learning change management and rapid reaching mature life cycle

It is one good question: Can the manager or CEO help whole organization to develop rapidly if he/she attempt to learn how to help his/her organization to implement different strategies to solve different challenges in different business life cycle stages? Does it easy to help the organization to grow up when a learning CEO or learning manager accepts to learn anything to compare a non learning manager in different business life cycle stages? Has it relationship between learning or non learning manager and rapid experiencing business life cycle stage and rapid developing business growth? I shall attempt to explain as below:

In fact, it is not essential to any managers or CEOs need to spend time to learn how any why what factors may influence their organizations to grow up to next business life cycle stage, but in comparison one learning how to change organizational life cycle stages manager and non-learning how to change organizational life cycle stages manger. Can learn attitude or strategy to help the manager to develop or expand his organization to next life cycle stage more easily or rapidly? I shall attempt to explain as below:

In fact, any organizations expect to change to next life cycle stage in success , can the manager(s) learn how to implement strategies to achieve to change management to their organizations‘ development in success? How the organizational management learns how to adapt organizational management change, it may be one important factor to influence whether the organization needs to spend how long time to reach growth life cycle stage from birth stage or reach mature life cycle stage from growth stage. So, it seems that how management spends time to learn how to change his/her organization. It will have relationship to the organization needs to spend long time to reach next life cycle stage successfully.

Hence, learning how to train employees in each life cycle stage, it is the important factor to influence any organizations succeed, the employee lifecycle is an ongoing process that starts and ends with competent employees in any managers' organizations. There are nine elements ofa successful change management process, if the organizational management expects whole organization can real reach to next life cycle stage in success. The nine elements of a successful change management process, any management needs to spend time to learn. They may include: readiness assessments, communication planning implementation, sponsor activities and sponsor roadmaps organizing, organizatons need to provide change management training for managers to learn how to achieve effectiveness as well as providing training development and delivery learning methods to them, resistance management learning and learning employee feedback and corrective action. Moreover, managements also need to spend time to learn change management steps in order solve any challenges in order to reach next life cycle stage easily.

The change management learning steps may include: Step 1: Urgency creation , step 2: Building every team serves to every department efficiently, learning how to create avision, how to communication of division, how to remove obstacles, going for quick wins, let the change mature, integrate the change. These elements are incorporated into change management phases process. For example, some elements of communication planning occur early in the lifecyle. At this stage, change management is not fully achieved effectively, so management needs to spend more time to learn how to achieve effective communication planning in order to achieve effective communication planning in order to keep whose organization employees can communicate to work efficiently. Also, it will help client service employees to know how to build good communication management method to deal or answer or satisfy their clients' sale service and improving service performance absolutely.

Because organizations are nor statis, they change , if one organization still stays long time in birth stage, it represents that the organization feels difficulties to continue develop . So, the management needs to find whether what challenges threaten its organization can not reach growth stage more easily. One failure changing management organization, it has these characteristics: failure to change, inexperienced management, not enough revenue, inadequate leadership. Hence, it has close relationship between employee life cycle and organizational life cycle . If the organizational management expects its organization can continue develop or reaches next life cycle stage in success, it needs to learn how to let employees to adapt when its organization is changing in order to keep efficience and improving service performance absolutely . So, I believe that it has relationship between learning change management and reaching to mature business cycle stage rapid and achieving long time staying in business cycle mature stage .

The question concerns that how management can learn to implement change management strategy in order to let his organization can reach mature cycle stage in short time as well as keep to stay in this mature life cycle stage in long time?

Firstly, we need to know what change management life cycle means ? For information technological industry example, it may be explained that the change management process is designed to help control of the life cycle of strategies, tactical and operational changes to IT services through standardized procedures. The goal of change managent is to control risk and minimize disruption to IT service and business operations. So, IT industry, the process change management maturity model presents five levels of organizational maturity in change management: The five level may include: from the lowest level 1 to the highest level 5, level 1: Absent or Ad hoc, level 2: Isolated projects , level 3: Multiple projects, level 4: Organizatinal standards and level 5: organizational competency. So, for IT , software manufacturing industry, if the management knows how to manage and change software manufacturing quality in order to satisfy manufacturing organization can follow software users' needs to change old function to new function and improve their qualities to achieve the highest level 5 organizational competency level.

Then, I believe that due to this organization's software management can learn how software user needs change and change its any kinds of software functions ( software life cycle), when its all softwares can be often changed to more new functions to create many different kinds of new software functions to satisfy software users needs and fight its software compettors in this often changing needs market. Due to software product may experience often changing life cycle stages. So, for often one learning software manager example, I believe that he can help this software organization to reach growth life cycle stage, even mature life cycle stage more easily in short time as well as he can also help his software organization to stay in mature life cycle stage long time if this software organizational manager can keep learning attitude to continue to create any new kinds of different functions software to satisfy software clients' changing needs for long time . Then, I believe that this software organization may experience or reach growth life cycle stage, even mature life cycle stage as well as continue staying long time on mature life cycle stage or avoid to encounter decline life cycle stage occurrence chance, if this software organization's softeare management can learn how to change software organization operation and software manufacture and sale strategy in order to satisfy this software users' needs in this software users' need often changing market . So, it is one example to explain why it has close relationship between learning organizational management method and business life cycle stages. As this software organization case, the software management needs often to create and change any new kinds of software functions in order to satisfy software users' needs . So, the software managers need to spend time to learn software life cycle stage , it can help the software organization may reach products life cycle stage, even mature life cycle stage in short time,even the software product organization may also stay long time in mature life cycle stage , when it can reach this stage. Hence, learning how to change organizational management or strategy, which is one important factor to help any organization can reach growth or mature life cycle stage eadily in short time.

As Lewin describes that the change as a three stage process of unfreezing, change and freezing . In this phases of change model, Lewin emphasizes that change is that a series of individual processes, but rather one that flows from one process to the next . So, in general, services mature firms pace greater emphasis on more bureaucratic form, control systems might need to change throughout the life cycle to fit in with. He explains they have relationship between both organizational life cycle stage and management control.

Effective management control may help the organization to reach mature life cycle in short time rapidly. So, leadership managment and the way of thinking are required to balance control and through several stages of growth, maturity , decline or re-grow changes in the external environment influence. Hence, managers position in each of the stages of life cycle and providing practical solutions are, however world where environment changes have proven a rapid growth, the management of varios , they also need to implement how to change their organizational cultures, strategies in order to let their organizations to reach mature stage with a distinction-oriented rapidly. Hence, to successfully implement change initiatives, for each phase of life cycle. Any organizations need to produce resistance to change ( the old model wins out over management boils down to improving the relationship ) learning the relationship between leadership style and the organization life cycle were important. The change from one organizational life cycle phase to another, it depends on how the manager‘c capacity to learn and change.

However, organizations at any stage of the life cycle are impacted by external environment, for example, threats in the start up stage differ from those in the maturity stage. So, managers must need often to learn when the right time is to be needed to change the goals, instead he also needs to learn types of changes in the maturity stage, comparisons with other, having strong personal and professional relationships in the organizaton's maturity stage. Hence, I believe that it has close relationship between learning change management and reaching maturity life cycle and staying long time in this stage.

● How to achieve the experience of mature life cycle reaching stage rapidly for product and service ?

Any businesses expect they can have chance or possibility to attempt to experience this nature life cycle stage, but it is not guarantee any kinds of businesses must may experience this the topest stage, the question is that: Have any methods may help any kinds of businesses to reach this the topest level of business life cycle, when their businesses had been developing or expanding in a period, e.g. after five years? So, it has no absolute to guarantee any kinds of businesses must may experience this the topest stage in one fixed time. How businesses can adapt to birth and growth life cycle stages in order to reach this the topest mature stage in their business life cycle stages? I shall attempt to explain whether it is possible that achieving what strategies may help businesses bring high successful chance to reach the business life cycle mature stage as below:

Product life cycle with maturity stage, it foucs as an important strategic inflection point. A number of techniques can help their businesses to attempt to reach this stage more easily. In fact, the product life cycle contains four distinct stages: introduction, growth, maturity, and decline. Each stage is associated with changes in the product's marketing position . Any firms can use various marketing strategies in each stage to try to proplong the life cycle of their products.

How do the firm extend the maturity stage of a product? I shall recommend change price , place or promotion extension strategy , what does change price extension strategies mean? Change prices mean proces can be lowered to allow ew customers to buy it as well as change place means that products can be sold in different countries or territories to gain more sales, change promotion means different advertising or sales promotion techniques can proplong the life of the product, giving it a new image. So, any organizations can attempt to achieve this extension strategies in order to adapt in different birth, growth and maturity stages for ther product sale easily. This extension strategies' characteristics is at the product;s price, sold places and promotin methods can be changed in order to adapt clients needs when their products are selling in birth, growth and maturity three stages in order to achieve the most effective sale effort and clients growth increasing for long time.

In fact, any product is like human beings, products also have a limited life-cycle and they pass through several stages in their life cycle. A typical product moves through five stages, namely, introduction or birth, growth, maturity or saturation and decline stages. So, when the product needs the maturity life cycle stage, in this maturity stage, it has these characteristics: The maturity stage of the product life cycle shows that sales will eventually peak and then slow down. During this stage, sales growth has started to slow down, and the product has already reached widespread acceptance in the market, in relative terms, utimately, during this stage, sales will peak . Hence, any businesses ought need to consider what key strategies can be implement to achieve the best sale performance throughout the different product life cycle stages and how to make the most of each stage. For example, when the product is selling in the birth stage, e.g. one author's book , his book is selling to the publisher in the first year, there are not many readers knew this book existence, so this book is not popular, its price ought not change high to compare similar topic book, e.g. story book in this year, but after this year, if there are many readers know this book and readers number can grow up rapidly. This author's this topic story book does not change, either increases or decreases , but its sale number has been significant increasing after the first year . So, this author's this story book ought be raised book price to attempt to sell easily. It is one good example of extension strategy to this author's this story book in its life cycle stages. So, such as ths publisher book sale case, it may attempt to achieve extension strategies to every author's book sale, it can follow every author's book prices, publishing places and promotion methods to help every author to sell in the most competitive book sale price, sale place choice and promotin methods in order to earn their readers growth aim . So, any book , it is as product to book shop, it will experience introduction, growth, and maturity life cycle stages. Some books may attract many readers to consider or some books may not attract many readers to consider to read . So, it causes their reading life cycle stages staying time will be different. So, extension strategies can help any books to be sold easily.

In fact, instead of product has life cycle stage, any service also has life cycle stage. There are five stages in service lifecycle. Thay may include: Service strategy, service design, service transition, service operation and continual servce improvement five stages. The service strategy phase of the service lifecycle provides guidance on how to design , develop and implement service management. Because any service business needs to manage to any employee service performance in order to provide excellent service quality, e.g. property management service to building tenants or property owners , if the peoperty management furm can train employees to provide excellent property management service to let their managing building clients to feel satisfactory. Then, the property management firm ought may keep long time property management service to this building. So, service provider will also experience service performance different stages.

In different service performance life cycle stages, such as this property management service case, they ought implement dfferent strategies in order to let their employees to know how to achieve service performance improvement to let their servicing building clients ( tenants or builgin owners) can feel their property management service can be continue improved to avoid to choose any property management service provider to replace it easily.

The purpose of the service strategy stage of the service life cycle is to define the perspective, position plans and pattern that a service provider needs to be able to execute to meet an organization business outomes. The objective of service strategy may include: An understandng of work strategy is thus either the concept of the product life cycle or the concept of the service life cycle is today at about to give a propsed new product or service , how and to what extent. This generally requires important changes in marketing strategies and methods, because any learning kinds of service or product lif cycle stage why and how to change to any organizational management, it may be an important tool for marketers, managers, and product and service providing designers alike, If specifies four individuals stages of a product's or service's life and offers guidance for developing strategies to make the best use of these stages and promote the overall success of the product or service in the marketplace.

Reasons managment needs to spend time to learn how to manage his/her product or service life cycle development stage? They may include: The product or service life cycle is determined by how long its marketable . Product or service life cycle also plays a critical role in marketing strategy . So, learning how to adapt your product or service to meet the coming trends , this is the stage what will occue in which differentiation when the kind of the product or service will have possible to reach the another new experience life cycle stage in order to adapt its business development more easily.

Hence, each stage is associated with changes in the product's or service's marketing postion . The organizational management can use various markting strategies in each stage to try to prolong the life cycle of your products or services . Any product or service reaches the marketplace, it enters the service or product life cycle . This product cycle typically has for stages: Introduction or birth, growth, maturity and decline ( and possibly deaths stages for product as well as service strategy stages includes service strategy. service design, servic transition, service operation, and continual service stages four service stages. So, the organization management can spend time to learn how to develop its business product or service needs to change in order to adapt marketing change in its product or service different life cycle stages. It can bring these benefits, such as: true benefits of product or srvice life cycle management may include, reduced time to makret, reduced market entry costs, more efficient and profitable distribution challen, higher return on investment from promotional cappaigns in possible, extending the lifetime of your product or service by adapting your approach as it moves through the lifecycle , for example, any management needs to learn what can make its products or services move from growth to maturity. After the introduction and growth stages, a product or service passes into the maturity stage. IN the first two stages , companies try to establish a market and then grow sales of their product or service to achieve as large , a share of that market as possible. Hence, marketers must be sure that a product or service has moved from one stage to the next before changing its marketing strategy. At each stage, marketing strategy varies. Strategy for the different stages of the product or service life cycle strategies may include: such as more benefits may be provided to the customers, e.g. extending the warranty period, guarantee period etc. However, company's market strategy depends on which stages the product or service is in its life cycle, for example, when one software manufacture company expects to expand its software sale market to overseas from local in growth stage. If it expects that it can reaches maturity stage in short time rapidly. It needs to implement technology innovation strategy for competition advantage reasons in global software sale markets development. Thus, the software organizational manager needs to spend time to learn what its present organizational characteristics are what resources and skills it owns or lacks, that gives it to comparative advantages over different countries to the operating changes that result in the learning curve to prepare this software product sale organizational maturity life cycle stage development more successfully. So, it needs to look at the advantages of focusing on what kinds of software manufacture and sale services in this software development industry whole life cycle stages and find the best or the most suitable competitive straregy, e.g. a discountinuous change to the software product development marketplace, what the global software product development industrial stage is and the tertiary or sale services sector durig the maturity life cycle stage to this softare manufacturer and sale organization strategy to this software firm during this growth stage may include example of it how changed its software product sales channels to which countries will be its another expanding sale market choice.

On conclusion, any organization management ought spend time to learn whether which strategies are the most suitable or the best to implement as well as how to implement when it is experiencing in the prodiuct or life cycle stage in order to spend less time to reach the maturity life cycle stage and proplong its maturity life cycle stage more success.

CHAPTER TEN

# Robotic future final mature life cycle development stage

Must Developed And Developing Countries Need Artificial Intelligent To Replace Human Job

Must developed and developing countries need artificial intelligent development? If one developed country, e.g. US, UK , Japan , Singapore it does not continue to develop artificial intelligence, robotic, then what disadvantges or weaknesses , it will encounter to compare when it chooses to continue to develop this artificial intelligent technology in society. If one developing country, e.g. China, Korea, Taiwan, it does not continue to develop artificial intelligence, robotic, then what disadantages or weaknesses, it will also encounter to compare when it chooses to continue to develop this artificial intelligent technology in in society. I shall explan the reasons why the results may cause to either the developed country, or the developing country as below:

- How AI help developing countries to communication and agriculture and learning and medical delivery development

Why can AI help developing countries ? Drones that pick inaccessible crops and mobile phones that give medical advice are two of the ways AI can transform life in the developing world. Artificial intelligence (AI) may improve the lives of the world's poor, the technology needed to revolutionise inefficient, ineffective food and healthcare systems in developing countries is well. For example, in low-income areas, agriculture and healthcare are two critical ecosystems that we can apply AI to immediately; this is not the far future, or even in five years.

Artificial intelligence (AI) has seeped into the daily lives of people in the developed world. From virtual assistants to recommendation engines, AI is in the news, our homes and offices. There is a lot of potential in terms of AI usage, especially in humanitarian areas. The impact could have a multiplier effect in developing countries, where resources are limited.

Emergency Response to developing countries' earthquake natural damage suddence occurrence predicting

AI and machine learning are still finding importance in emerging markets, but certain applications have emerged and are now widely used. For instance, predictive models for disaster relief enable first responders to automatically analyze large-scale behavior and movement through multiple sources of data including social media platforms, web forums, news sources, etc. Based on collected data, responders can scale reconstruction efforts and distribute supplies in a timely manner.

Why and how AI can assist farmers to predict when the earthquake occurs suddenly in order to avoid or reduce the natural damage to their agriculture productive number loss. For example, In 2015, when a major earthquake hit Nepal, more than 8 million people were affected. During the aftermath, drones were used to map and assess the destruction and speed up the rescue mission. The town of Sankhu, situated about 20 kilometers northeast of Kathmandu, was among the highly affected locations. In May 2018, my company Fusemachines and GeoSpatial Systems partnered with Sankhu's city officials to use drones and artificial intelligence in an effort to automatically estimate the reconstruction need. After processing data accumulated from a drone-powered aerial mapping of the region, the team fed this data to advanced machine learning algorithms. Combining drone imagery, digital mapping and machine learning, the team configured region modeling and infrastructure development with higher accuracy. Another organization known as One Concern, a California-based startup, has created a predictive AI program called Seismic Concern to accurately predict seism and is also working on solutions for wildfires, floods and hurricanes.

Smart AI Agriculture

Another application of AI in developing countries is smart agriculture. Farmers monitor crops more effectively and make better predictions on planting, weeding and harvesting using AI tools. It can also be used to analyze one plant at a time and add pesticides only to infected plants and trees instead of spraying pesticides across large swaths of crops. One California-based tech company is an example of this use of AI. So, the developing countries farmers in rural parts of India are also using AI to increase yields through better access to information about the farming season than they would normally have. Technology-enabled process automation offers the agribusiness industry the chance for remarkable growth -- not only in developed countries but around the world. There's a unique opportunity to increase yields, cut down labor costs and improve people's health.

Medicine Delivery to developing countries' patients urgent need

Companies are also leveraging AI to improve access to health care in some of the most remote areas of the world. In Rwanda, for example, Zipline is using drones to deliver medical supplies and blood to hospitals and clinics that are difficult to access by car. This has dramatically impacted people living in remote parts of the country because they are able to get medical help when needed. The drone system in Rwanda has also helped reduce waste of blood by 95%, as noted by Zipline. One Concern has created an AI program called Seismic Concern that accurately predicts seismic events and is also working on solutions for floods, wildfires and hurricanes. The medical field may actually benefit the most from emerging technologies in developing countries.

Assistance to reduce teaching work workload or psychological pressure to teachers in developing countries' schools

Another vital area benefiting from innovative technologies like AI is education. Advanced technologies can enhance how we learn, teach and perform tasks. In most developing countries, schools lack experienced teachers and resources to enhance students' knowledge. As a result, many students still have to walk long distances to get to the nearest school, which has created education gaps, especially in rural areas. AI tools such as personalized learning assistants can simplify learning by making tutoring services and learning materials accessible to all students, wherever they are. Machines can be automated to help students learn basic concepts without a tutor, which companies like Carnegie Learning are working on. This would allow students to learn at any time from anywhere. With AI, education is made easy and accessible to more people.

The initial usage of AI in developing countries has been at a micro level -- solving small, specific problems in a defined industry. As machine learning advances and there is a higher utilization of AI, we will see more complex issues being targeted and resolved. When duly adopted, AI can positively impact future developing countries people everyday lives not just in disaster intervention, education, health care and agriculture but can also help in mitigating poverty, malnutrition and pollution. Especially, in developing nations, to leverage AI's true potential and create a snowball effect. Startups are defining a holistic and humanitarian approach to building more sophisticated, AI-ready societies. Stakeholders in the AI landscape should understand the strengths and nuances of the developing world as well as the limitations of AI and create localized solutions and applications.

Why does smart phone help developing countries communication ?
Internet Seen as Positive Influence on Education but Negative on Morality in Emerging and Developing Nations. Internet access differs substantially across the 32 emerging and developing countries polled, with the lowest rates of internet use in South Asian and sub-Saharan African nations. Within countries, computer owners, young people, the well-educated, the wealthy and those with English language ability are much more likely to access the internet than their counterparts. To access the internet, people increasingly use smartphones rather than more cumbersome fixed landline connections and computers. Around the world, both smartphones and basic-feature phones alike are used for sending messages and taking pictures.

In fact, many developing countries young people, students are popular to use smart phones for internet usage aim, instead of communication. Moreover, many developing countries working people are also popular to use smart phones for any working usage in their working time , even non working time any time. So, smart phones (AI) phones will be important communication or leisure tools to developing countries people in the future. Unless, it is one day, scientists can develop another new communication tool to replace smart phones. So, artificial intelligence will be important to influence developing countries people , how to improve or bring positive learning attitudes to students in their daily learnnng lifes. as well as how to raise developing countries people, how to raise working people efficiency or improve performace in their daily working lifes. So, AI may bring positive learning or working attitudes to developing countries working people and students both.

The Positive Impact of Mass Media in Developing Countries
Radio, newspapers, television, Internet, social media, etc., all of these are forms of mass media. Each of these outlets has the capability of bringing information to thousands of people with one device. While in some communities it is easy to take advantage of these communication outlets such as television and Internet access, not everyone has access to such outlets. Radio is one of the most common forms of mass media in

developing countries because it's affordable and uses less electricity than many other forms of mass media, but only approximately 75 percent of people in developing countries have access to a radio, and roughly 77 percent of people in rural areas have access to electricity.

For developing countries that have implemented forms of mass media in their communities, there have been numerous positive outcomes are influenced to impact developing countries mass media by artificial intelligence as below:

When AI is participated to developing countries mass media, it can influence any radio, television audiences raise more attention to each other through social media platforms such as Facebook and Twitter and create, organize and initiate street protests and campaigns. Furthermore, having access to social media in developing countries, people are able to connect to those that they usually wouldn't have the chance to talk to. Moreover, AI Provides educational opportunities- In many countries, the division between local and national languages as well as issues of literacy can make communication difficult. With the use of mass media, a bridge can be built between these two gaps. In India, there is a radio station that provides information in local languages and respects local culture and traditions. One of the main ways is to create public awareness of what is going on with businesses and government officials. The media plays an important role in giving people the opportunity to act against injustice, oppression and misdeeds that they otherwise wouldn't know about. Information on available healthcare, a mass radio broadcast was sent out encouraging parents to seek treatment at local healthcare facilities for their sick children. With this mass outreach on healthcare, the encouragement of people to take their children to healthcare facilities saved thousands of lives. This easy way of encouraging others and bringing awareness about certain diseases was made possible through a simple radio broadcast. Finally, when AI is particiapted to media, it may bring many social issues to life that otherwise would remain unknown to many people. In developing countries and communities like Burkina Faso, when the radio broadcast was released about malaria, diarrhea and pneumonia, people were educated and moved to action and knew to take their children to healthcare facilities for preventative care. As it is seen, having access to different media outlets is vital for those in developing countries. Here are three ways that those in developing countries can implement mass media to help their people and communities.

When AI is participated to any internet radio or internet newspaper mass online listening or reading channel. It can provide online radios or newspapers in public places- By providing online radios and newspapers in public areas it gives community members to access news, information and emergency warnings. Even though radios can be on the cheaper side, there are still many people that can't afford to have a radio in their home. By providing one in a local place, not only would it better educate the community members but also it will bring the community together. So, it can make media outlets a two-way platform- Creating a two-way platform between the community and those who are behind the radio stations, newspapers or broadcasts makes the community feel involved and that their voices are being heard. An organization called Soul City in sub-Saharan Africa is showing how well two-way platforms work by engaging their listeners and having them contribute thoughts and ideas about complex issues. Because developing countries radio listening audiences or newspaper readers are popular to accept computer online radio listening channel or online newspaper reading channel to replace traditional paper newspapers or radio machines. So, AI may raise their listening news or reading news leisure feeling from online mass media channel in the future.

● Why do developed countries need to develop AI

Artificial intelligence, or AI, is driving massive shifts across the globe, and every day more questions arise. What impact will AI have on the workforce and how can we prepare for it? How can we encourage economy-boosting and job-creating technologies? How can we ensure that AI will be implemented ethically and with minimal bias? How will society benefit? For developed country, such as US example. None of the US, Israel and Russia have a formal national AI policy yet. Private sector companies such as Google, Amazon and Apple and the US department of defence are driving the bulk of AI investment in the United States. Though Israel does not have a specific policy, it is keenly focused on AI and has seen the number of AI start-ups triple since 2014.

Developed country may learn whether what weakness it is lacking when it does not continue to develop AI from one another developed country. Which countries are approaching AI most effectively, and to what degree is there opportunity for greater international collaboration? It may be too early to tell; however, when analyzing the best practices of existing national AI policies, there is much that can be learned. These are the specific areas

to consider. When one developed country continue to develop or research AI, it may bring these benefits as below:
On gathering Data aspect, from self-driving vehicles to smart cities, data is the driver behind AI. Innovation in the United States is limited without a national strategy that answers questions about protocol and ownership. France and Denmark, on the other hand, are opening government data. France is hosting troves of centrally collected public and private data that it plans to make available as part of its strategy. Conversely, by taking a restrictive position on issues of data collection (as indicated by the implementation of General Data Protection Regulation), the EU is putting manufacturers and software designers at a disadvantage while balancing the demand for privacy. On raising technologica talent aspect, the demand for AI talent far outweighs the available supply. As a result, almost every nation's strategy addresses talent development. Canada's AI strategy is distinct in that it primarily focuses on research and talent strategy. The country boasts AI degree programmes and is building a $127 million research facility in Toronto. Companies like Facebook and my own company, Uptake, are investing in Canada to access this talent pool. On AI legal technological innovation aspect, a whole host of legal questions swirl around AI. The country is developing a bill for AI liability that will be ready in March 2019. The government hopes the legal framework will attract investors by providing a simple, comprehensive guideline to enable the broad use of AI systems. So, when the developed country applied AI technology to assist any lawyers to work, then AI can help them to reduce the workload to draft any legal documents more easier. So, any developed countries lawyers' draft legal documents time must reduce if the developed countries lawyers accept to apply AI to assist their legal works. One of the great promises of AI is its potential for improving quality of life. But without the right planning and oversight, we risk exacerbating problems of inequality or marginalizing groups of people. As an example, India's AI strategy is focused on leveraging the technology not only for economic growth, but also for social inclusion.

AI may bring what benefits to developed countries

From SIRI to self-driving cars, artificial intelligence (AI) is progressing rapidly. While science fiction often portrays AI as robots with human-like characteristics, AI can encompass anything from Google's search algorithms to IBM's Watson to autonomous weapons. Artificial intelligence today is properly known as narrow AI (or weak AI), in that it is designed to perform a narrow task (e.g. only facial recognition or only internet searches or only driving a car). However, the long-term goal of many researchers is to create general AI (AGI or strong AI). While narrow AI may outperform humans at whatever its specific task is, like playing chess or solving equations, AGI would outperform humans at nearly every cognitive task.

Why research AI safety? Would AI bring war when AI is continued to develop by developed countries? In the near term, the goal of keeping AI's impact on society beneficial motivates research in many areas, from economics and law to technical topics such as verification, validity, security and control. Whereas it may be little more than a minor nuisance if your laptop crashes or gets hacked, it becomes all the more important that an AI system does what you want it to do if it controls your car, your airplane, your pacemaker, your automated trading system or your power grid. Another short-term challenge is preventing a devastating arms race in lethal autonomous weapons.

In the long term, an important question is what will happen if the quest for strong AI succeeds and an AI system becomes better than humans at all cognitive tasks. As pointed out by I.J. Good in 1965, designing smarter AI systems is itself a cognitive task. Such a system could potentially undergo recursive self-improvement, triggering an intelligence explosion leaving human intellect far behind. By inventing revolutionary new technologies, such a superintelligence might help us eradicate war, disease, and poverty, and so the creation of strong AI might be the biggest event in human history. Some experts have expressed concern, though, that it might also be the last, unless we learn to align the goals of the AI with ours before it becomes superintelligent.

There are some who question whether strong AI will ever be achieved, and others who insist that the creation of superintelligent AI is guaranteed to be beneficial. At FLI we recognize both of these possibilities, but also recognize the potential for an artificial intelligence system to intentionally or unintentionally cause great harm. We believe research today will help us better prepare for and prevent such potentially negative consequences in the future, thus enjoying the benefits of AI while avoiding pitfalls.

How can AI be dangerous when developed countries continue to develop AI to become weapon to replace soldiers?

Most researchers agree that a superintelligent AI is unlikely to exhibit human emotions like love or hate, and that there is no reason to expect AI to become intentionally benevolent or malevolent. Instead, when considering

how AI might become a risk, experts think two scenarios most likely:
The AI is programmed to do something devastating: Autonomous weapons are artificial intelligence systems that are programmed to kill. In the hands of the wrong person, these weapons could easily cause mass casualties. Moreover, an AI arms race could inadvertently lead to an AI war that also results in mass casualties. To avoid being thwarted by the enemy, these weapons would be designed to be extremely difficult to simply "turn off," so humans could plausibly lose control of such a situation. This risk is one that's present even with narrow AI, but grows as levels of AI intelligence and autonomy increase.
The AI is programmed to do something beneficial, but it develops a destructive method for achieving its goal: This can happen whenever we fail to fully align the AI's goals with ours, which is strikingly difficult. If you ask an obedient intelligent car to take you to the airport as fast as possible, it might get you there chased by helicopters and covered in vomit, doing not what you wanted but literally what you asked for. If a superintelligent system is tasked with a ambitious geoengineering project, it might wreak havoc with our ecosystem as a side effect, and view human attempts to stop it as a threat to be met. So, a super-intelligent AI will be extremely good at accomplishing its goals, and if those goals aren't aligned with ours, we have a problem. You're probably not an evil ant-hater who steps on ants out of malice, but if you're in charge of a hydroelectric green energy project and there's an anthill in the region to be flooded, too bad for the ants. A key goal of AI safety research is to never place humanity in the position of those ants.
Why the recent interest in AI safety ?
Stephen Hawking, Elon Musk, Steve Wozniak, Bill Gates, and many other big names in science and technology have recently expressed concern in the media and via open letters about the risks posed by AI, joined by many leading AI researchers. The idea that the quest for strong AI would ultimately succeed was long thought of as science fiction, centuries or more away. However, thanks to recent breakthroughs, many AI milestones, which experts viewed as decades away merely five years ago, have now been reached, making many experts take seriously the possibility of superintelligence in our lifetime. While some experts still guess that human-level AI is centuries away, most AI researches at the 2015 Puerto Rico Conference guessed that it would happen before 2060. Since it may take decades to complete the required safety research, it is prudent to start

it now.

Because AI has the potential to become more intelligent than any human, we have no surprise way of predicting how it will behave. We can't use past technological developments as much of a basis because we've never created anything that has the ability to, wittingly or unwittingly, outsmart us. The best example of what we could face may be our own evolution. People now control the planet, not because we're the strongest, fastest or biggest, but because we're the smartest. If we're no longer the smartest, are we assured to remain in control?

A captivating conversation is taking place about the future of artificial intelligence and what it will/should mean for humanity. There are fascinating controversies where the world's leading experts disagree, such as: AI's future impact on the job market; if/when human-level AI will be developed; whether this will lead to an intelligence explosion; and whether this is something we should welcome or fear. But there are also many examples of of boring pseudo-controversies caused by people misunderstanding and talking past each other. When one developed country continue to develop AI, can itself country's all factories workers will lose jobs, due to AI can replace them to do simple works in factories, or any public transport drivers, e.g. bus drivers, ferry , tram, train drivers, they will lose jobs, when AI ( non manual driving drivers ) can replace all public transport drivers. So, some occupations will lose if developed countries continue to develop or research AI to replace human to do some simple jobs, such as some cooking jobs can be done by AI. So, it is possible that future cookers won't be needed, because AI cooking skills may be better than them to cook any good taste chinese or western food in restaurants. If you drive down the road, you have a subjective experience of colors, sounds, etc. But does a self-driving car have a subjective experience? Does it feel like anything at all to be a self-driving car? Although this mystery of consciousness is interesting in its own right, it's irrelevant to AI risk. If you get struck by a driverless car, it makes no difference to you whether it subjectively feels conscious. In the same way, what will affect us humans is what superintelligent AI does, not how it subjectively feels.

In fact, AI may be make any brokers jobs in financial market. the main concern of the beneficial-AI movement isn't with robots but with intelligence itself: specifically, intelligence whose goals are misaligned with ours. To cause us trouble, such misaligned superhuman intelligence needs no robotic body, merely an internet connection – this may enable outsmarting financial markets, out-inventing human researchers, out-manipulating human leaders, and developing weapons we cannot even understand. Even if building robots were physically impossible, a super-intelligent and super-wealthy AI could easily pay or manipulate many humans to unwittingly do its bidding. So, future brokers will be replaced by AI, when AI can be made to own financial brokers' analytical mind to make more accurate whether the share price will rise up or fall down to compare human financial brokers' analytical mind. The robot misconception is related to the myth that machines can't control humans. Intelligence enables control: humans control tigers not because we are stronger, but because we are smarter. This means that if we cede our position as smartest on our planet, it's possible that we might also cede control.

Not wasting time on the above-mentioned misconceptions lets us focus on true and interesting controversies where even the experts disagree. What sort of future do you want? Should we develop lethal autonomous weapons? What would you like to happen with job automation? What career advice would you give today's kids? Do you prefer new jobs replacing the old ones, or a jobless society where everyone enjoys a life of leisure and machine-produced wealth? Further down the road, would you like us to create superintelligent life and spread it through our cosmos? Will we control intelligent machines or will they control us? Will intelligent machines replace us, coexist with us, or merge with us? What will it mean to be human in the age of artificial intelligence?

Why do developed countries people need AI ?

Why do we assume that AI will require more and more physical space and more power when human intelligence continuously manages to miniaturize and reduce power consumption of its devices. How low the power needs and how small will the machines be by the time quantum computing becomes reality? Why do we assume that AI will exist as independent machines? If so, and the AI is able to improve its Intelligence by reprogramming itself, will machines driven by slower processors feel threatened, not by mere stupid humans, but by machines with faster processors? What would drive machines to reproduce themselves when there is no biological incentive, pressure or need to do so?

Who says superior AI will need or want to have a physical existence when an immaterial AI could evolve and preserve itself better from external dangers. What will happen if AI developed by competing ideologies, liberalism vs communism, reach maturity at the same time, will they fight for hegemony by trying to destroy each other physically and/or virtually. If AI is programmed to believe in God, and competing AI emerges programmed by muslims, christians or jews, how are the different AI's going to make sense of the different religious beliefs, are we going to have AI religious wars? What if the "powers that be" greatest fear is the emergence of a super AI that police's and rationalizes the distribution of wealth and food. A friendly super AI that is programmed to help humanity by, enforcing the declaration of Human Rights (the US is the only industrialized country that to this day has not signed this declaration) ending corruption and racism and protecting the environment.Most benefits of civilization stem from intelligence, so how can we enhance these benefits with artificial intelligence without being replaced on the job market and perhaps altogether?

Key to the process of machine learning are neural networks. These are brain-inspired networks of interconnected layers of algorithms, called neurons, that feed data into each other, and which can be trained to carry out specific tasks by modifying the importance attributed to input data as it passes between the layers. During training of these neural networks, the weights attached to different inputs will continue to be varied until the output from the neural network is very close to what is desired, at which point the network will have 'learned' how to carry out a particular task. A subset of machine learning is deep learning, where neural networks are expanded into sprawling networks with a huge number of layers that are trained using massive amounts of data. It is these deep neural networks that have fuelled the current leap forward in the ability of computers to carry out task like speech recognition and computer vision.

In conclusion, when developed countries continue to develop AI, it may bring positive advantages to bring raising productivies, or efficiencies, but it may also raise unemployment ratio to any low skill or low knowledge jobs in ther societies. However, human future society will need to change to be better to raise our living standard. But AI is one kind the best choice tool to achieve this aim in our future, so I agree developed countries continue to develop or research AI to be the super -human machine.

Reference

A. Castano et. al. " Automatic detection of dust devils and clouds at Mars" Machine vision and applications, Oct. 2008, vol. 19, no 5-6, pp. 467-482.

Accenture, " Why artificial intelligence is the future of growth"(2017) <http://www.accenture.com/us-en/insight-a rtificial-intelligence-future-

growth>.

D. Schedidt , Unmanned Air Vehicle Command And Control, Handbook Of Unmanned Air Vehicles, Springer-Verlag, 2014. Facebook (AI) Research Available at https://research.facebook.com/ai, research at google, machine intelligence available at http://research.google.com/pubs/machineintellige nce.html; micro soft research-machine learning and artificial intelligence available at http://research.microsoft.com/en-us/research- areas/machine-learning-ai.aspx.

International Federation Of Robotics, 2016. IFR press release world robotics report. IFR, org . 29 Sept. Accessed Feb. 01, 2017. http://www.ifr.org/news/ifr-press-release/world-robitics report -2016-8321.

K, Fedra , "GIS and environmental modelling" in environmental modelling with GIS, edited by M.F. Goodchild.B.O. Parks and L.T. Steyaert, Oxford University press, pp. 35-50, 1994.

Keynes, J.M. (1933). Economic possibilities for our grandchildren (1930). Essays in persuasion, pp.358-73.

Mckinsey & Company (2013, May). Disruptive technologies: Advices that will transform life, business and the global economy , USA.

Ministry of economy, trade and industry, Japan, 2015, Japan's robot strategy. Ministry of economy, trade and industry.

Ray Kurzweil , The age of spiritual machines (1999) is cited numerously through this chapter: Kurzweilai.net http://www.kurzweilai.net

Rich, Elaine & Knight, Kevin, Artificial Intelligence Second Edition, 1991, New York; Mc-Graw-Hill.

Artificial intelligence bank service working environment

Focus on outcomes not technology.Artificial Intelligence: Waiting to be unleashed? The Insider Column - When Digital Transformation misses. Are you meeting the demands of the new digital consumer? Will your legacy mindset compromise your digital competitiveness? Can artificial intelligence create online remote office new business service market in global ?

The Future of Artificial Intelligence In The Workplace:

Is AI going to displace workers or come as a benefit to them?

Is AI going to displace workers or come as a benefit to them? Getty
Smart technologies aren't just changing our homes; they're edging their way into their numerous industries and are disrupting the workplace. Artificial Intelligence (AI) has the potential to improve productivity, efficiency and accuracy across an organization – but is this entirely beneficial? Many fear that the rise of AI will lead to machines and robots replacing human workers and view this progression in technology as threat rather than a tool to better ourselves.
With AI continuing to be a prominent online office service business to replace human actual office working environment, businesses need to realize that self-learning and black-box capabilities are not the panacea. Many organisations are already beginning to see the incredible capabilities of AI, using these advantages to enhance human intelligence and gain real value from their data. As there is increasing evidence demonstrating the benefits of intelligent systems, more decision-makers in the boardroom are gaining a better understanding of what AI can really offer. Research conducted by EY explains "organizations enabling AI at the enterprise level are increasing operational efficiency, making faster, more informed decisions and innovating new products and services." Can articial intelligent technology create remote office working environment to replace our traditional actual office work environment ? Can we do not need to go to office to work , when any office staffs ,e.g. managers, clerk, etc. they can apply artificial intelligent technology and online technology to work at home, such as remote office working environment ?
The first companies employing AI systems across the board will gain competitive advantage, reduce cost of operations and remove head counts. Whilst this may be a positive from a business perspective, it is obvious why this a worry for those working in roles at risk of displacement. The introduction of these technologies will likely trigger an issue with unions and job security due to the substantial operational changes. Although AI will affect every sector in some way, not every job is at equal risk. PwC predicts a relatively low displacement of jobs (around 3%) in the first wave of automation, but this could dramatically increase up to 30% by the mid-2030's. Occupations within the transport industry could potentially be at much greater risk, whereas jobs requiring social, emotional and literary abilities are at the lowest risk of displacement.

A positive future with artificial intelligence to bring remote online office working environment chance:

Many businesses and individuals are optimistic that this AI-driven shift in the workplace will result in more jobs being created than lost. As we develop innovative technologies, AI will have a positive impact on our economy by creating jobs that require the skill set to implement new systems. 80% of respondents in the EY survey said it was the lack of these skills that was the biggest challenge when employing AI programs.

It is likely that artificial intelligence will soon replace jobs involving repetitive or basic problem-solving tasks, and even go beyond current human capability. AI systems will be making decisions instead of humans in industrial settings, customer service roles and within financial institutions. Automated decisioning will be responsible for tasks such as approving loans, deciding whether a customer should be onboarded or identifying corruption and financial crime.

Organisations will benefit from an increase in productivity as a result of greater automation, meaning more revenue will generated. This thus provides additional money to spend on supporting jobs in the services sector.Due to the vast array of jobs that could be impacted by AI, it is fundamental to address the potential pitfalls of these technologies. Business need to overcome the trust and bias issues surrounding AI by achieving an effective and successful implementation that makes it possible for everyone to benefit.

Governments must ensure that gains from AI are shared widely across society to prevent social inequality between those affected and unaffected by these developments. For example, this could be through increased investment into training. With the additional cost-savings from implementing AI systems, employers should also focus on upskilling their current employees.

To properly leverage the power of AI, we need to address the issue at an educational level, as well as in business. Education systems needs to focus on training students in roles directly associated to working with AI, including programmers and data analysts. This requires more emphasis to be put on STEM subjects (science, technology, engineering and mathematics). Also, subjects centered around building creative, social and emotional skills should be encouraged. Whilst artificial intelligence will be more productive than human workers for repetitive tasks, humans will always outperform machines in jobs requiring relationship-building and imagination. Hence, artificial intelligence will change our world both inside and outside the workplace. Instead of focusing on the fear surrounding automation, businesses need to embrace these new technologies to ensure they implement the most effective AI systems to enhance and compliment human intelligence.

Artificial Intelligence (AI) in Banking working environment

Artificial Intelligence (AI) is a fast-evolving technology, gaining popularity all around the world. Several industries have already adopted AI for various applications, getting better and smarter day by day. In the past few years, the banking sector has also become one of the leading adopters of Artificial Intelligence. Most banks and financial institutions are implementing AI to add more efficiency to their back-office and lessen security risks.

As per Statista, the AI market in the United States is forecasted to reach 7.35 billion U.S. dollars in 2018. Some major applications of AI include classification, image recognition, object identification, and automated geophysical feature detection. Speaking of banking and financial institutions, JPMorgan Chase, Wells Fargo, Bank of America, CitiBank, and other leading U.S. banks have already implemented AI in their systems, helping consumers manage their daily banking needs more efficiently.

AI technology can bring better Customer Support in bank service environment

Several pieces of evidence advocate that the customers willingly prefer self-service options which allow them to chat with a virtual assistant as if it were a live customer representative. Most leading banks have already added virtual assistants to their instant website chatbots, voice response systems, and mobile applications. Artificial Intelligence considers each interaction as a teachable moment, so the chatbots (virtual assistants) keeps getting better while understanding customers. With AI, virtual assistants can deliver better customer support. It also allows sentiment analysis, so the virtual assistant can determine when individuals are getting frustrated and instantly transfer them to a live agent.

Enhanced Banking Services

AI streamlines the banking process while giving customer service a new level of comfortability. It allows banks to meet customers' expectations with comprehensive digital support. With Artificial Intelligence, you can achieve greater precision and accuracy. From cash transfer to bills payment, cards management, and other support, AI can significantly enrich the satisfaction level of your customers. All of these operations can be easily managed through desktops, smartphones, and other mobile devices.

Scam Recognition

With an immense growth of banking fraud, scam recognition and reduction has become challenging for the banking sector. Several banks tried to identify the factors and powerful solutions but couldn't succeed. However, AI makes it easier to detect the factors involved in frauds and support investigators. It improves financial security with advanced fraud prevention tactics. Artificial Intelligence works as a real-time scam solution for the banking sector while handling complex situations and tactics. Based on advanced data crunching, AI can detect fraud by flagging unusual transactions. It also feeds back into the consumer's profile which subsequently builds a secure environment.

Advanced Data Analytics

One of the main advantages of AI is its ability to complete tedious tasks through intricate automation, resulting in better productivity. Based on a machine learning algorithm, AI can quickly consume and process a massive amount of data at an expedited level. The enormous speed brings efficiency to financial services, providing scope for personalized offerings to consumers. What's even more, AI makes faster decisions while carrying out actions quickly. With such advantages, it is nearly obvious that the majority of banks and financial institutions will adopt AI to stay competitive and deliver better customer support. However, several cons are also associated with a machine learning algorithm. As it continues to learn and grow, the decision-making capabilities may create problems in the near future.

Disadvanages of AI in Banking Sector

Artificial intelligence is also expected to massively disrupt banks and traditional financial services. Some of its disadvantages are listed below.

Highly Expensive

Production and maintenance of artificial intelligence demand huge costs since they are very complex machines. AI also consists of advanced software programs which require regular updates to meet the needs of the changing environment. In the case of critical failures, the procedure to reinstate the system and recover lost codes may require enormous time and cost.

Bad Calls

Though Artificial Intelligence can learn and improve, it still can't make judgment calls. Humans can take individual circumstances and judgment calls into account when making decisions, something that AI might never be able to do. Replacing adaptive human behavior with AI may cause irrational behavior within ecosystems of humans and things.

Distribution of Power

There is a constant fear of AI superseding or taking over the humans. Artificial intelligence can give a lot of power to the few individuals who are controlling it. Hence, AI carries the risk and takes control away from humans while dehumanizing actions in several ways.

Unemployment

Replacement of the workforce with machines can lead to wide-reaching unemployment. Moreover, if the use of AI becomes rampant, people will be highly dependent on the machines and lose their creative power. Unemployment is a socially undesirable issue. Individuals with nothing to do can lead to the devastating use of their minds. Be it banking or any other sector; Artificial intelligence can effectively increase the unemployment rate.

Artificial Intelligence delivered to wrong hands can turn out to be a serious threat to humankind. If individuals start thinking destructively, they can generate havoc with these advanced machines. The challenges introduced by the emergence of artificial intelligence revolve around several things. However, AI is a right balance of skill and emotions which is continually growing. Artificial intelligence provides banks, financial institutions, and tech companies with significant competitive advantages. Nevertheless, it can completely transform the financial sector and make it faster, but this will only be possible if the financial industry can manage the security risk of systems based on AI.

What does artificial intelligence mean for the bank service office workers?

With all these new artificial intelligence use cases comes the question of whether machines will force humans into obsolescence. The jury is still out: Some experts vehemently deny that artificial intelligence will automate so many jobs that millions of people find themselves unemployed, while other experts see it as a pressing problem.

"The structure of the workforce is changing, but I don’t think artificial intelligence is essentially replacing jobs in bank service working environment. It allows us to really create a knowledge-based economy and leverage that to create better automation for a better form of life. It might be a little bit theoretical, but I think if you have to worry about artificial intelligence and robots replacing some bank service jobs, e.g. bank security, bank enquiry service,. But, AI can not replace bank counter service staffs to do saving or withdrawing money transfer tasks when any customers prepare to save money or withdraw money in bank counters. As this technology develops, the AI bank service will see new startups, numerous saving or withdraw transactions from consumer won't be raise more easily.

AI to Banking and Finance industry

The banking and finance industry plays a major role in our lives. I mean the world runs on money and banks are essentially the gatekeepers that regulate that flow. Did you know that the banking and finance industry heavily relies on artificial intelligence for things like customer service, fraud protection, investment, and more? A simple example is the automated emails that you receive from banks whenever you do an out of the ordinary transaction. Well, that's AI watching over your account and trying to warn you of any fraud.

AI is also being trained to look at large samples of fraud data and find a pattern so that you can be warned before it happens to you. Also, when you hitch a little snag and chat with bank's customer service, chances are that you are chatting with an AI bot. Even the big players in the finance industry use AI to analyze data to find the best avenues to invest money so they can get the most returns with the least risk. That's not all, AI is poised to play an even bigger role in the industry as major banks across the world are investing billions of dollars in the AI technology and we all will observe its effects sooner than later

How AI influences our daily working life in any office working environment

Can AI bring only disadvantages? If AI can bring disadvantges, what are its disadvantages to any working environment ?The entire tech world is debating the consequences of artificial intelligence and the part AI is going to play in shaping our future. While we might think that artificial intelligence is at least a few years away from causing any considerable effects on our lives, the fact remains that it is already having an enormous impact on us. Artificial intelligence is affecting our decisions and our lifestyles every day. Don't believe me? I shall indicate some product examples how AI anticipates which can influence our working culture in any office environment.

Examples of how Artificial Intelligence assistance to office working environment may include as below:

1. Smartphones

Smartphones have become the most indispensable tech product that we own today and we use it almost all the time. Well, if you are using a smartphone, you are interacting with AI whether you know it or not. From the obvious AI features such as the built-in smart assistants to not so obvious ones such as the portrait mode in the camera, AI is impacting our lives in every day office working environment.

In fact, the two examples that I provided that our working world of AI and how it is effecting our working lives. Firstly, there are the obvious AI elements which most of us have some knowledge about. For example, when you are using a smart assistant in office, whether it's Google Assistant, Alexa, Siri, or Bixby, you more or less know that these assistants are based on AI. However, when we are using a feature such as the portrait mode effect while shooting a picture, we never consider that AI might be behind that too. Have you ever thought how the Google Pixel phones or iPhones can capture such great portrait shots? The answer is artificial intelligence. So, when any office workers need to find any knowledge to solve their working problem immediately in any offices. They may apply AI smart phone tools to help them to apply online channel to search any new knowledge to attempt to solve their working problem in possible, when their computers have none any computers in offices.

Now more and more manufacturers are including AI in their smartphones with big chip manufacturers including Qualcomm and Huawei producing chips with built-in AI capabilities. The AI integration is helping in bringing features like scene detection, mixed and virtual reality elements, and more. AI is going to play an even major role in the coming years. We are already seeing the huge emphasis on AI with the latest Android and iOS updates. Features like app actions, splices, and adaptive battery in Android Pie and Siri shortcut and Siri suggestions in iOS 12 are made possible with AI. So, next time if any office workers think AI is not effecting them, take out your smartphone to replace computers to find any knowledge to help you to solve any tasks problems immediately in offices.

2. Social Media Feeds

If you are thinking that smart cars don't personally effect you as they are still not in your country or city, well, how about something which you use on a daily basis. Even if you are living under a rock, there's a high probability that you are tweeting from underneath it. If Twitter's not your choice of poison, maybe it's Facebook or Instagram, or Snapchat or any of the myriad of social media apps out there. Well, if you are using social media, most of your decisions are being impacted by artificial intelligence. So, any office workers may apply AI to help them to gather any new information to solve any difficult task problems , if their managers can not assist them to solve any sudden tasks problem, they are encountering to need to solve any working complex tasks problem internet social media in any any office working environment immediately.

From the feeds that office staffs can see in their working timeline to the notifications that you receive from these apps, everything is curated by AI. AI takes all your past behavior, web searches, interactions, and everything else that you do when you are on these websites and tailors the experience just for you. The sole purpose of AI here is to make the apps so addictive that you come back to them again and again, and I am ready to place a bet that AI is winning this war against you.

3. Online Ads Network

One of the biggest users of artificial intelligence is the online ad industry which uses AI to not only track user statistics but also serve us ads based on those statistics. Without AI, the online ad industry will just fail as it would show random ads to users with no connection to their preferences what so ever. AI has become so successful in determining our interests and serving us ads that the global digital ad industry has crossed 250 billion US dollars with the industry projected to cross the 300 billion mark in 2019. So next time when any product developers are going online and seeing ads or product recommendation, know that AI is impacting to any new products advertisement method more efficiently.

4. AI can be any office security

While we can all debate the ethics of using a broad surveillance system, there's no denying the fact that it is being used and AI is playing a big part in that. It is not possible for humans to keep monitoring multiple monitors with feeds from hundreds if not thousands of cameras at the same time, and hence, using AI makes perfect sense. With technologies like object recognition and facial recognition getting better and better every day, it won't be long when all the security camera feeds are being monitored by an AI and not a human. While there's still time before AI can be fully implemented such as security in any offices, this is going to be our future.

5. Smart Keyboard Apps

Smart Keyboard Apps. Granted, not everyone loves dealing with on-screen keyboards. However, they have become far more intuitive, allowing users to type comfortably and faster. What has probably proved to be a catalyst for them is the integration of AI. The smart keyboard apps keep a tab on the writing style of a user and predict words and emojis accordingly. Thus, typing on the touchscreen has become faster and more convenient. Not to mention, artificial intelligence also plays a vital role in pin-pointing misspellings and typos. So, any office workers can apply smart keyboard apps to help their to raise typing efficiency and reduce wrong typing word in error when they need to type any document in offices.

6. E-Commerce

` AI-driven algorithms have kind of given the much-needed impetus to e-commerce to provide a more personalized experience. According to several reports, its usage has vastly increased sales and also played a good part in building loyal relationships with customers. Thus, companies take advantage of AI to deploy chatbots to collect pivotal data and also predict purchases to create a customer-centric experience. Yet to come across this shift of strategy? Just spend some time with sites like Amazon and eBay and you will soon get to know how fast the landscape is changing around you – for the better! So, Ai can help any businesses to achieve e-commerce sale channel more easily.

7. Smart Email Apps

In any office working environment, if you still find your inbox cluttered with too many unwanted messages, chances are pretty high that you are still stuck with an old school email app. You heard it right! Modern email apps like Spark make the most of AI to get rid of spam messages and also categorize emails so that you can quickly access the important ones. What's more, they also offer smart replies based on the messages you receive to help you reply to any email quickly. The "Smart Reply" feature of Gmail is a great example of this. It uses AI to scan the text of the email and provides you with contextual answers. So, AI can help any office staffs to know who had sent any message from email and respond their email immediate , when AI can help any offices to avoid to receive any email spam rubbish email message in any time, even after working hours, it means that AI is working to help any office staffs to avoid to receive any email spam rubblish message in any time. So, when they go to office to work, even they go home after working hours. They can know whether what the important email messages are sent to their office email in boxes any time. Then, they can send email to respond their customers‘ enquires any time. So, AI can help any office workers can have chance to work at homes.

The Future of Artificial Intelligence In The Workplace

Smart technologies aren't just changing our homes; they're edging their way into their numerous industries and are disrupting the workplace. Artificial Intelligence (AI) has the potential to improve productivity, efficiency and accuracy across an organization – but is this entirely beneficial? Many fear that the rise of AI will lead to machines and robots replacing human workers and view this progression in technology as threat rather than a tool to better ourselves.

With AI continuing to be a prominent buzzword in 2019, businesses need to realize that self-learning and black-box capabilities are not the panacea. Many organisations are already beginning to see the incredible capabilities of AI, using these advantages to enhance human intelligence and gain real value from their data. As there is increasing evidence demonstrating the benefits of intelligent systems, more decision-makers in the boardroom are gaining a better understanding of what AI can really offer. Research conducted by EY explains "organizations enabling AI at the enterprise level are increasing operational efficiency, making faster, more informed decisions and innovating new products and services."

Today In: Cybersecurity

The first companies employing AI systems across the board will gain competitive advantage, reduce cost of operations and remove head counts. Whilst this may be a positive from a business perspective, it is obvious why this a worry for those working in roles at risk of displacement. The introduction of these technologies will likely trigger an issue with unions and job security due to the substantial operational changes. Although AI will affect every sector in some way, not every job is at equal risk. PwC predicts a relatively low displacement of jobs (around 3%) in the first wave of automation, but this could dramatically increase up to 30% by the mid-2030's. Occupations within the transport industry could potentially be at much greater risk, whereas jobs requiring social, emotional and literary abilities are at the lowest risk of displacement.

A positive future with artificial intelligence

Many businesses and individuals are optimistic that this AI-driven shift in the workplace will result in more jobs being created than lost. As we develop innovative technologies, AI will have a positive impact on our economy by creating jobs that require the skill set to implement new systems. 80% of respondents in the EY survey said it was the lack of these skills that was the biggest challenge when employing AI programs. It is likely that artificial intelligence will soon replace jobs involving repetitive or basic problem-solving tasks, and even go beyond current human capability. AI systems will be making decisions instead of humans in industrial settings, customer service roles and within financial institutions. Automated decisioning will be responsible for tasks such as approving loans, deciding whether a customer should be onboarded or identifying corruption and financial crime. Organisations will benefit from an increase in productivity as a result of greater automation, meaning more revenue will generated. This thus provides additional money to spend on supporting jobs in the services sector.

How to take advantage of AI to any offices

Due to the vast array of jobs that could be impacted by AI, it is fundamental to address the potential pitfalls of these technologies. Business need to overcome the trust and bias issues surrounding AI by achieving an effective and successful implementation that makes it possible for everyone to benefit. Governments must ensure that gains from AI are shared widely across society to prevent social inequality between those affected and unaffected by these developments. For example, this could be through increased investment into training.With the additional cost-savings from implementing AI systems, employers should also focus on upskilling their current employees.

To properly leverage the power of AI, we need to address the issue at an educational level, as well as in business. Education systems needs to focus on training students in roles directly associated to working with AI, including programmers and data analysts. This requires more emphasis to be put on STEM subjects (science, technology, engineering and mathematics). Also, subjects centered around building creative, social and emotional skills should be encouraged. Whilst artificial intelligence will be more productive than human workers for repetitive tasks, humans will always outperform machines in jobs requiring relationship-building and imagination. Artificial intelligence will change our world both inside and outside the workplace. Instead of focusing on the fear surrounding automation, businesses need to embrace these new technologies to ensure they implement the most effective AI systems to enhance and compliment human intelligence

How AI can help office workers to do tasks more easily

Companies are currently spending big on artificial intelligence and machine learning initiatives to the tune of $12 billion, but estimates put that figure as high as $57.6 billion by 2021, according to the International Data

Corporation (IDC). With such massive shifts, the focus is usually on what we might lose, but it shouldn't be. A recent report on the future of work from the McKinsey Global Institute suggests that while only about 5% of jobs can be completely eliminated by automation, the rise of AI requires workers to beef up both technical and soft skills in order to stay competitive.

What's seldom discussed is how AI can revolutionize our jobs. It's now possible to pinpoint peak productivity for a single day, improve communication in meetings (even before people ever work together face to face), or even teach you to be a better leader, all thanks to AI platforms. I shall indicate these advantages to bring any office benefits from AI assistance as below:

1. AI can help any companies to get better to hire the best applicants

AI has the greatest potential to change the way companies find candidates, according to Alexander Rinke, cofounder and CEO of Celonis. The company's process-mining technology helps businesses to understand the areas where automation can help humans, he says. In HR departments, Celonis can help identify how fast workers come and go, the cost per hire, and which positions take the longest to fill. AI helped enable one customer's ability to identify bottlenecks in recruitment and reduced process costs internally by 30% as well as get them hired more quickly, he says.

Crafting a resume has never been easier, nor has landing an interview. Another example is how recruitment software provider iCIMS, in partnership with Google, is helping job seekers find jobs directly through the search engine, thanks to Google's AI and machine learning capabilities. Susan Vitale, iCIMS's chief marketing officer says that in addition to reducing the number of expired job postings, machine learning is underlying a private beta program of Google's Cloud Jobs Discovery model. "For a candidate searching for, say, a CTO role, Cloud Job Discovery will serve up CTO positions as well as jobs with titles that are similar, but not verbatim, such as chief technology officer or chief technical officer," says Vitale. This model also allows for conceptual search results, such as serving up job listings for cashiers, sales associates, and store associates when someone searches for one versus just only showing jobs that exactly match the keyword search criteria, she adds.

2. AI can help any office workers to raise much more productive efficiencies

John Furneaux, CEO and cofounder of Hive, says predictive analytics will help us better understand how we work. "It can tell us just about everything we want to know about teams and collaboration, for example, if men or women get more done in the afternoon, and if summer Fridays are a myth," he says. (Everyone thinks summer Fridays aren't productive, but in reality there's no difference between those and other Fridays during the year–productivity is equally low.)

Using a data set of over 30,000 completed actions across Hive workspaces, Furneaux says they were able to identify some notable trends in productivity. For example, men were far more productive early in the day, with a sharp decline in the afternoon, while women had a slower start to the day but were far more productive in later hours than their male counterparts. And analyzing chat messages revealed that women appear to complete more tasks when chatting, suggesting they use communication as a key tool to completing work. Similarly, Nintex Hawkeye analyzes data on business processes by types, users, roles, and departments to see who's doing the work and how long it takes them to do it. Management can monitor and analyze those metrics in real time.

3. AI can help any managers to make the most fair compensation and eliminate wage gaps to every staffs

Tanya Jansen, cofounder of the compensation management platform beqom, says that AI and predictive analytics can eliminate unconscious bias from compensation. Jansen says that AI based on a variety of rules including education, experience, certifications, and more can make compensation more fair and help businesses move closer to closing pay gaps. "Specifically, AI can help solve gender pay gaps and the CEO-to-worker pay gap, in which pay ratios of Fortune 500 companies range from 2:1 at the low end to nearly 5000:1 at the high end," she says. Additionally, the use of AI-driven compensation technology to make pay more fair can mitigate the risk of employee turnover, which costs businesses as much as 33% of a worker's annual salary to replace them.

4. AI can help any office staffs to arrange better meetings

Augmented Reality (AR) is still in its infancy, but AI and machine learning are the core components that make it work. As such, Christa Manning, the vice president and solution provider research leader at Bersin, Deloitte Consulting LLP, says that AR can help workers find the right information, in the right place, at the right time to make the best decisions wherever they may be working. For example, as more companies adopt video meetings and collaborative workspaces, it's likely we'll begin to see HR-curated information like talent profiles and work styles layered over interactions through AR."Imagine being in a video conference with a colleague and having direct insight into their communication style, seeing tips on how to best interact with them or reminders of what needs to be discussed. SO, AI can help any organizations to conclude or find the best methods to solve any problems after their every discussion in any meetings.

How AI is improving onboarding and training. AI coaching tools first learn by observing how different employees conduct specific tasks. Then these tools can walk new employees through how to complete those tasks—or even coach existing employees on how to do things more effectively or efficiently. Chorus is a great example of this technology. It analyzes sales calls while they happen, offering tips to help sales reps manage the cadence of meetings and use the most effective messaging. It also records all sales calls and compiles statistics for each sales rep, providing everyone with the tools they need to help them close more deals and conduct more effective calls. Another example is Cogito, a tool that combines AI with behavioral science to help customer service employees provide better phone support. It monitors calls for voice signals, providing real-time suggestions to representatives on how to improve the conversation.

5. AI can help any managers to be better leaders

Indiggo, a platform powered by a proprietary AI tool called "indi," functions as a brain that has consumed all the knowledge the company has gathered in its 15 years of operation. It also uses an algorithm to provide an estimate of how much time is wasted by a company by analyzing the size of its management team. Then it taps their calendars to see how they spend their time, and walks individual managers through a type of Q&A to make sure they are clear on what their top three priorities are, and how that relates to the organization's priorities, which will indicate if that strategy is moving forward or not. "The counterintuitive impact of these advances is that they actually make human work truly irreplaceable," Alexander Rinke, the cofounder and CEO of Celonis says. As such, he reminds us, "Humans are much better at processes that involve reasoning, judgment, and interaction with people." So, AI can recommend more accurate and useful opinions to help any managers to solve their managing challenges in office any time.

How AI is eliminating repetitive administrative tasks

There are a lot of tasks that knowledge workers spend time on that provide little—if any—value.For example, say you need to schedule a meeting to get consensus on a decision before moving forward, but you need five people to join the meeting. It's easy to spend a ton of time sending email back-and-forth or finding an open slot on everyone's calendar.That's not the most rewarding use of your time for you or your company.Tools like X.ai give employees AI-powered personal assistants that perform administrative tasks like scheduling, rescheduling, and cancelling meetings.

How AI is transforming internal communications and support

Personnel on the teams that provide employee support have their hands full with other responsibilities, too. HR teams work on building the kind of company people love working for. IT maintains the company's network and keeps data secure. Office managers frequently run big events like holiday parties.These tasks are crucial, but they're often hard for teams to focus on because they're busy answering routine questions. AI service desks like askSpoke allow employee support teams to balance their service commitments with other important responsibilities by reducing interruptions from rote, repetitive requests.Employees can askSpoke for whatever they need over Slack, email, SMS, and the web. askSpoke's friendly AI will automatically provide a prompt response.

How AI is transforming marketing, sales, and customer service

AI-powered chatbots help with external support as well. Just like with internal support tools like askSpoke, these chatbots learn from real marketers, salespeople, and customer service reps and are eventually able to answer questions as accurately as a knowledgeable person.For example, chatbot for Messenger helps customers plan their vacations. It books flights, hotels, and cars, highlights destination attractions, and even provides answers to questions like "Where can I go for $100 expense budget only?"

How AI is transforming business data and analytics

It's hard to run a competitive business today without data. But even massive amounts of data are useless without a way to transform that data into valuable insights. That's typically why you'd want to hire a data scientist—which just happens to be one of the most difficult roles to fill. How AI is fighting fraud and transforming security. Have you ever taken a call from your bank to find that someone used your debit card fraudulently? Most likely, your bank used some form of AI to detect the fraudulent transaction and decline it. Applying the same basic technology to the workplace helps identify security risks and keeps customer, employee, and company data safe. AI-powered software can automatically detect and address threats among thousands or millions of signals that humans would never be able to parse (especially not in real-time).

How AI is transforming productivity
While AI is transforming the workplace in many different ways across every industry, it's impacting productivity most of all. When your office staffs don't have to scroll through calendars to look for open meeting times, build reports in spreadsheets to look for insights, or spend your day answering the same questions over and over again, you're more productive. Workers are freed from redundant and mindless tasks, giving them more time to do work that matters, solve problems, and exercise their creativity. Some tools use AI to specifically monitor and boost productivity. For example, Deloitte's LaborWise provides company leaders and managers with productivity analytics that help them identify areas where labor costs are too high, impediments that slow people down, and departments that need additional staff.
In conclusion, what AI means for the workplace of the future. While some will dramatize the negative impacts of AI, cognitive computing, and robotics, these powerful tools will also help create new jobs, boost productivity, and allow workers to focus on the human aspects of work. Essentially, automation frees companies and their employees up to be more empathetic, to focus on things like the customer experience, employee engagement, and workplace culture.

What are traditional office tools to be replaced by AI ?
Artificial intelligence (AI) is predicted to eliminate over a million jobs in the next few years, potentially replacing lower level positions like administrative assistants with humanoid robots or voice assistants. But in the nearer future, fresh AI-driven software and products are also moving to eliminate non-human elements of the workplace by replacing traditional office tools, including both physical products and everyday electronic processes. Why should businesses switch from the tried-and-true to emerging technology? Many of the experts TechRepublic talked to said the AI options streamline business practices, making their adopters work smarter instead of harder. I shall indicate these office tools ,they can be applied to help any office staffs to finish their these tasks in office, they may include as below:

1. Scheduling
Workloud's end-to-end, cloud-based workforce management software takes scheduling from paper or Excel and moves it to the cloud. Everything from clocking in and out to monitoring employee absences is fully digitalized.Schedules and timesheets are accurate, created easily, and

accessible through the service's web, tablet, and mobile apps. The software can also be used for absence management.

2. Employee talent selection

Using AI and organizational behavior science, can be used to replace internal spreadsheets and databases designed to monitor human capital. By mining employee attributes and experiences, the software can recommend who would be best for a project. The software also collects reviews after projects to better predict successful employee-project matches.The traditional hiring process is slow, biased and inaccurate, By removing humans from the beginning stages of the process, it can become faster and more fair, and result in better hires.

AI software automates the hiring process, using online simulations instead of manual screenings and interviews. Using the software, employers can include tasks in a job application, allowing job candidates to show technical skills that may be necessary for a job. Employers can't rule out candidates until they see how the candidate performs, eliminating bias that occurs in the resume reading stage. Both sides also automatically receive updates about each other's steps, reducing the amount of time it takes to .

3.Timesheets: Allocate

Using AI and machine learning, the software registers an employee's computer activity throughout the day. The data, which can also pull information from email and calendars, is used to suggest timesheet entries to reflect a more accurate amount of time an employee spent working. The employee can review and revise as necessary. However, the software doesn't spy on or monitor employees. The data is only available to each employee, while others in the company can only see the timesheet's output, which Allocate said would be the same information available if a manual sheet was used. So,replacing manual timesheets with Allocate has three advantages: More accurate time entry, project analytics, and "'unsucking' the work experience."

4. Document storage

By using AI to read and analyze business and legal documents, AI can store all of the important document-based information in the cloud. The severe reduction in print-outs means less paper and ink, fewer products like binder clips and boxes to store and organize all of the paper, and more employee time freed up from not needing to manually sort through every document.

For example, in any lawyer offices, legal professionals' morale in the industry can suffer when they are pushed into performing such dull, repetitive tasks like sorting through and coding documents by hand, With AI tools to automate those duties, lawyers can focus on more meaningful projects and boost the business's and clients' success as a result. While focused on law firms, businesses that have a lot of unstructured data in documents may also be able to use the service to free up employee time and save on printing costs.

5. Scanners: Adobe Scan

While documents are moving to the cloud more and more, sometimes a physical copy of a document still needs to be scanned using a bulky office scanner. Adobe Scan, an app that condenses a scanner to the size of a smartphone, can rid offices of the need for an in-house scanner. Users can download and open the app, then hold their device over whatever they need to scan. Adobe Sensei then turns the scan into a PDF, and sends it to the Adobe Document Cloud. The app can transform any image into digital text that can then be searched and used electronically. The app streamlines the scanning process, making scans cleaner and more immediate. For businesses already using Adobe services, the app makes documents easily accessible.

6. Landline phones

While landlines in homes are increasingly less common, the same cannot be said for offices. But using chatbots and AI integrations, RingCentral is trying to replace traditional office landline phone systems. The platform offers over 100 integrations, including that AI landline phones can let employees check their voicemail, and a Gong.io option that listens to call recordings to find traits of successful employees than can be used in training. An add-on for Gmail lets users switch from emailing back and forth to a voice session without needing to look up contact information. AI landline phone is easy to adopt and use in the workplace, and is more customizable than standard phone systems, said David Lee, vice president of platform products. Compared to the traditional option, the cloud-based option is "future-proof.

How artificial intelligence can raise office efficiency

Artificial Intelligence is already impacting every industry through automation and machine learning, bringing concerns that AI is on the fast track to replacing many jobs. But these fears aren't new, says Dan Jackson, director of Enterprise Technology at Crestron, a company that designs workplace technology. "I'd argue this is no different than when we moved from an agricultural to an industrial economy at the turn of the last century. The percentage of people working in agriculture significantly decreased, and it was a big shift, but we still have plenty of jobs 100 years later," he says. Anytime society experiences a major technological advancement, we need to be prepared for it to change the way we live and work. It's hard to imagine what the future of jobs will look like with AI, but that future exists. And optimists suggest that, like the sewing machine to the textile industry, AI will make us better, more efficient and faster workers.

In fact, many experts agree that AI has the potential to eliminate mundane, administrative work, while we will always rely on human workers to be empathetic, collaborative, creative and strategic. But it's impact on any industry lies in the hands of the business leaders who are responsible for adopting AI strategies.

● Training presents challenges

A recent study of 1,000 global companies by Accenture found that AI is already creating three new categories of jobs: trainers, explainers and sustainers. Trainers are the people who teach AI systems how to act -- whether it's language, human behavior or the intricacies of human interaction. Explainers are the liaison between technology and business leaders, providing more insight and clarity into machine learning for the non-tech workers. Sustainers are the workers required to maintain AI systems and troubleshoot any potential issues. Some jobs were highly technical and required advanced degrees, but other roles demanded innately human things such as empathy and interaction. Downstream jobs, such as those in sales, marketing, or service will change to take advantage of the insights from AI, but many of the core skills will remain. However, it might sound like any job related to AI will require years of technical knowledge, but that isn't the case. We've already seen a shift in tech hiring -- companies often need highly specific skill sets that are hard to find in potential candidates. As a result, more businesses are hiring employees with the right soft skills, and then training them in technical skills.

An office effort measured approach to AI

The real takeaway is that any approach to AI will need to consider the human aspect of every business. AI has great potential to increase efficiency and accuracy and it's already been proven in certain industries. For example, the use of AI In banking to identify and money laundering schemes. It's also improved healthcare by "increasing the speed and accuracy" of cancer diagnosistics. AI can also help reduce the cost and length of human trafficking investigations, a situation where time is precious. In these examples, AI hasn't replaced jobs, but has positively

impacted efficiency.
Thus, we need to ensure our education system responds to equip young people with the appropriate skills and adaptability, while businesses and public organizations must invest in training. Perhaps most of all, we need to encourage imagination and willingness to experiment. The organizations that can innovate with AI will reap the benefits. Their growth will make them the primary source of future jobs. Companies have a choice when implementing AI. They can choose to effectively implement systems that make employee's lives easier and find creative ways to leverage the technology. It's up to employers to ease fears for workers around AI and build strategies that benefit everyone. Hence, some AI experts believe AI can only raise efficiency to some office tasks, however, AI can not still raise efficiency to all office tasks for any office deparments. The reasons are because some office tasks which can only dominate to finish by human office workers. These office tasks are as below:
How can leaders and managers improve employee productivity while still saving time? These below tasks, AI experts ensure that AI can not help any office workers to raise their efficiencies as below:

1. Office managers can not delegate to AI to help them to do. While this tip might seem the most obvious, it is often the most difficult to put into practice. We get it–your company is your baby, so you want to have a direct hand in everything that goes on with it. While there is nothing wrong with prioritizing quality (it is what makes a business successful, after all), checking over every small detail yourself rather than delegating can waste everyone's valuable time. Instead, give responsibilities to qualified employees, and trust that they will perform the tasks well. This gives your employees the opportunity to gain skills and leadership experience that will ultimately benefit your company. You hired them for a reason, now give them a chance to prove you right.

2. Office managers can not match Tasks to Skills to AI. Knowing your employees' skills and behavioral styles is essential for maximizing efficiency. For example, an extroverted, creative, out-of-the-box thinker is probably a great person to pitch ideas to clients. However, they might struggle if they are given a more rule-intensive, detail-oriented task. Asking your employees to be great at everything just isn't efficient–instead, before giving an employee an assignment, ask yourself: is this the person best suited to perform this task? If not, find someone else whose skills and styles match your needs.

3. Office managers can not teach AI to replace them how to communicate and teach their low level staffs how to work effectively. Every manager knows that communication is the key to a productive workforce. Technology has allowed us to contact each other with the mere click of a button (or should we say, tap of a touch screen)–this naturally means that current communication methods are as efficient as possible, right? Not necessarily. A McKinsey study found that emails can take up nearly 28% of an employee's time. In fact, email was revealed to be the second most time-consuming activity for workers (after their job-specific tasks). Instead of relying solely on email, try social networking tools (such as Slack) designed for even quicker team communication. You can also encourage your employees to occasionally adopt a more antiquated form of contact...voice-to-voice communication. Having a quick meeting or phone call can settle a matter that might have taken hours of back-and-forth emails. All of above communication tasks, I believe that AI can not do better than managers in offices.

4. AI can not keep Goals Clear and focused to be better than managers. You can't expect employees to be efficient if they don't have a focused goal to aim for. If a goal is not clearly defined and actually achievable, employees will be less productive. So, try to make sure employees' assignments are as clear and narrow as possible. Let them know exactly what you expect of them, and tell them specifically what impact this assignment will have. One way to do this is to make sure your goals are "SMART" – specific, measurable, attainable, realistic, and timely. Before assigning an employee a task, ask yourself if it fits each of these requirements. If not, ask yourself how the task can be tweaked to help your workers stay focused and efficient.

5. AI can not know how to incentivize Employees to work more efficiently. One of the best ways to encourage employees to be more efficient is to actually give them a reason to do so. Recognizing your workers for a job well done will make them feel appreciated and encourage them to continue increasing their productivity. When deciding how to reward efficient employees, make sure you take into account their individual needs or preferences. For example, one employee might appreciate public recognition, while another would prefer a private "thank you." In addition to simple words of gratitude, here are a few incentives managers can know

how to incentivize their staffs to work efficiently, but AI is only one machine, it can not perform very good.

6. AI does not know how to assist managers to train and Develop employees. Reducing training, or cutting it all together, might seem like a good way to save company time and money (learning on the job is said to be an effective way to train, after all). However, this could ultimately backfire. Forcing employees to learn their jobs on the fly can be extremely inefficient.
So, instead of having workers haphazardly trying to accomplish a task with zero guidance, take the extra day to teach them the necessary skills to do their job. This way, they can set about accomplishing their tasks on their own, and your time won't be wasted down the road answering simple questions or correcting errors. Past their original training, encourage continued employee development. Helping them expand their skillsets will build a much more advanced workforce, which will benefit your company in the long run. There are a number of ways you can support employee development: individual coaching, workshops, courses, seminars, shadowing or mentoring, or even just increasing their responsibilities. Offering these opportunities will give employees additional skills that allow them to improve their efficiency and productivity. But, AI do not know how to improve any office workers' performance more easily than managers.

- How can AI be dangerous to office working environment?

Most researchers agree that a superintelligent AI is unlikely to exhibit human emotions like love or hate, and that there is no reason to expect AI to become intentionally benevolent or malevolent. Instead, when considering how AI might become a risk to any office working environments, experts think two scenarios most likely:
The AI is programmed to do something devastating: Autonomous weapons are artificial intelligence systems that are programmed to kill. In the hands of the wrong person, these weapons could easily cause mass casualties. Moreover, an AI arms race could inadvertently lead to an AI war that also results in mass casualties. To avoid being thwarted by the enemy, these weapons would be designed to be extremely difficult to simply "turn off," so humans could plausibly lose control of such a situation. This risk is one that's present even with narrow AI, but grows as levels of AI intelligence and autonomy increase. So, if some businessmen apply AI to be business weapon to attack or steal their business competitors' business secret, e.g. contract document, employee performance report, profit report, even

business secret document. Then, AI will be one business competitor weapon more than business assistant role in any business market. So, whether AI is office assistant or business competitor weapon, it depends on how the businessmen apply them to assist their business development.

The AI is programmed to do something beneficial, but it develops a destructive method for achieving its goal: This can happen whenever we fail to fully align the AI's goals with ours, which is strikingly difficult. If you ask an obedient intelligent car to take you to the airport as fast as possible, it might get you there chased by helicopters and covered in vomit, doing not what you wanted but literally what you asked for. If a superintelligent system is tasked with a ambitious geoengineering project, it might wreak havoc with our ecosystem as a side effect, and view human attempts to stop it as a threat to be met.

As these examples illustrate, the concern about advanced AI isn't malevolence but competence. A super-intelligent AI will be extremely good at accomplishing its goals, and if those goals aren't aligned with ours, we have a problem. You're probably not an evil ant-hater who steps on ants out of malice, but if you're in charge of a hydroelectric green energy project and there's an anthill in the region to be flooded, too bad for the ants. A key goal of AI safety research is to never place humanity in the position of those ants. Because AI has the potential to become more intelligent than any human, we have no surefire way of predicting how it will behave. We can't use past technological developments as much of a basis because we've never created anything that has the ability to, wittingly or unwittingly, outsmart us. The best example of what we could face may be our own evolution. People now control the planet, not because we're the strongest, fastest or biggest, but because we're the smartest. If we're no longer the smartest, are we assured to remain in control?

A captivating conversation is taking place about the future of artificial intelligence and what it will/should mean for humanity. There are fascinating controversies where the world's leading experts disagree, such as: AI's future impact on the job market; if/when human-level AI will be developed; whether this will lead to an intelligence explosion; and whether this is something we should welcome or fear. But there are also many examples of of boring pseudo-controversies caused by people misunderstanding and talking past each other. To help ourselves focus on the interesting controversies and open questions — and not on the misunderstandings — let's clear up some of the most common myths.

There have been a number of surveys asking AI researchers how many years from now they think we'll have human-level AI with at least 50% probability. All these surveys have the same conclusion: the world's leading experts disagree, so we simply don't know. For example, in such a poll of the AI researchers at the 2015 Puerto Rico AI conference, the average (median) answer was by year 2045, but some researchers guessed hundreds of years or more. There's also a related myth that people who worry about AI think it's only a few years away. In fact, most people on record worrying about superhuman AI guess it's still at least decades away. But they argue that as long as we're not 100% sure that it won't happen this century, it's smart to start safety research now to prepare for the eventuality. Many of the safety problems associated with human-level AI are so hard that they may take decades to solve. So, any businessmen ought have business moralty to know whether they ought how to apply their AI to assist their business development in our future office environment to be more moral.

- Five ways to use AI to improve business efficiency to these office tasks

Regardless of a company's size or type, its executives typically look for ways to help it operate as efficiently as possible. They understand the link between efficiency and profitability. If employees waste too much time with drawn-out processes or complicated tasks, it'll be hard for the enterprise to remain profitable and adapt to challenges. Fortunately, artificial intelligence (AI) supports the need for effective business operations. Here are five ways enterprises can use AI for help: 5 ways to use AI to improve business efficiency image.Getting the best results from AI means looking at where bottlenecks exist, then figuring out if and how it might remove or minimise them. AI can help any offices to improve or raise efficiency to these tasks aspects as below:

1. Use AI to answer queries and support customer engagement

Chatbots are an increasingly popular option for businesses to try, and they use AI to work. Companies often build chatbots that can answer any questions from customers that come through outside of business hours. Some identify the nature of a person's problem, then either attempt to tackle it with preprogrammed answers or pass the communications to a human support worker. The retail industry, in particular, saw success by deploying chatbots. Global data collected by Juniper Research shows an estimated 2.6 billion retail-based chatbot interactions in 2019, and the company forecasts the number to rise to 22 billion in 2023.

Chatbots are excellent for answering simple questions like "How late are you open today?" or "Do you have gluten-free menu options?" Getting quick answers to queries like those increases the chances customers will choose to do business with one company over another. Equally importantly, when chatbots can give responses in a matter of seconds, there's no need for humans to stop what they're doing and address the questions.

2. To enhance reporting speed and accuracy

Company reports reveal things such as which products are selling the fastest and where they're most popular. They can also confirm the impacts of marketing campaigns on product sales, break down the costs of a new packaging choice or shipping method, and much more. However, as anyone that files reports knows, creating them is a painstaking task, and trying to rush through the process could cause mistakes. Some forward-thinking companies are combining AI with big data analytics. Doing this brings better forecasts and takes some of the burdens off the people who prepare the reports. AI also helps conquer the inevitability of mistakes. Even the most careful people make blunders, often because of mental fatigue.

AI learns to spot patterns in data and gets smarter with time. This means reports get finished faster and contain more-reliable information. The reliability aspect is crucial, especially since recently published research indicated two-thirds of the senior executives polled had no confidence or trust in big data. Using AI does not mean companies can do without data scientists. However, depending on the technology allows them to reduce the uncertainty that may otherwise exist. It also prevents employees who work with a company's data from being asked to recheck the findings, even if they initially took appropriate precautions to ensure accuracy.

3. To improve data transfer speeds

Fast data transfers help AI technology work. Concerning some information-intensive applications like virtual reality (VR), any slow transmissions greatly interfere with the realism, and content immersion people should enjoy after strapping on a VR headset. As it turns out, AI can improve data transfer speeds, too. For example, services exist that boost speeds across any wide-area network (WAN). Users enjoy consistently accelerated rates regardless of the kind of information transferred. Some companies have solutions that can reduce WAN job times by up to 98%. These AI-driven options work particularly well when companies need to move information between data centres or cloud environments.

4. To assist the IT team with identifying genuine cyberthreats and anomalies

One of the ongoing challenges faced by IT teams of all sizes is to separate the true cyber threats from false alarms. The difficulties associated with categorising the two types may mean cybersecurity professionals waste time getting to the bottom of things that are ultimately nonissues. They might miss the actual threats that could derail a company's operations. Besides detecting possible intrusions associated with a network, AI can screen for software abnormalities that may make it easier for cybercriminals to orchestrate their attacks successfully. It can also find malicious software hackers installed. Due to this kind of information and the advantages of receiving it through real-time updates, IT security teams can work more productively. They can use the majority of their resources on the threats that matter most to the company's stability.

Some organisations have even used AI to help them conquer the substantial skills shortage in the cybersecurity industry. At Texas A&M University, the Security Operations Center deals with about a million attempted hacks each month. The facility has some full-time workers, but students comprise most of the staff. They work alongside AI that aids in threat monitoring, detection and remediation. Before students see possible threats, the smart technology finds and groups them. This approach saves time and lets the team get to work investigating the problems and deciding how to handle them.

5. To streamline the time-to-hire metric when filling new positions

Statistics show the average time required to hire a person for an open position ranges from 12.7 to 49 days, depending on the industry. The timing also varies based on the type of work a job requires. For example, it takes a shorter amount of time overall to find someone for an administrative or human resources position than one associated with a creative or advertising role. Then, of course, interviews are more extensive for high-profile work.

Human resources professionals increasingly use AI to cut down on the time between first posting a job and finding the ideal individual to hire. For example, an AI platform could look for particular desired keywords in submitted resumes, saving hiring managers from poring over the documents themselves. AI can also pitch in during interviews. A company called VCV recently raised $1.7m to further develop its AI tool that has voice and facial recognition components. Candidates are asked to record videos of them answering interview questions, but they can't prepare for the specific content in advance.

In conclusion, AI Can Boost Efficiency at All Types of Companies. The examples here highlight why so many company leaders conclude that if they use AI, they could cut down on inefficiencies. Getting the best results from AI means looking at where bottlenecks exist, then figuring out if and how it might remove or minimise them. But, AI still lack enough effort to help all staffs to raise efficiency to all department tasks in any office environments.

● How the office energy Department is using AI to solve some of their office staffs electricity toughest challenges in their office working environment.

Insights from artifical intelligence has the potential to transform nearly every aspect of the world as we know it. Today, it is being applied to accelerate the pace of discovery in a wide variety of areas including energy, materials science, health care, national security, emergency response, transportation, and more. AI can be trained to help any energy department to gather data to avoid energy waste to be used to any organizations. So, AI is such as one super machine to do more accurate judgement to help any energy scientists to find the best methods to help any organizations to avoid to waste to use any energy daily. Then, organizations can avoid to spend too much energy to use in offices and they can save more money and avoid energy shortage challenge causes more easily. When the office managers can apply AI ability to reason and put it into a more automated format in a computer system to their every staffs' computer and record their computer electricity use record in their offices every day.

How can AI help offices to save energy or avoid to waste energy ?

The next industrial revolution is already happening. Artificial intelligence (AI) is ushering in an era of technologies that are faster, more adaptable, more efficient, and making the world more digitally connected. AI is best described as complementary to human intelligence, delivering the computing power to crunch numbers too big for people and recognize patterns too tedious for the human eye. In a Harvard Business Review study of 1,500 companies, it was found that the most significant performance improvements were made when humans and machines worked together. As AI becomes one of society's greatest assets, it's especially helpful for solving problems that seem larger than life — like protecting our natural environment.

Through machine learning, robotics, drones, and the internet of things (IoT), society can achieve better monitoring, understanding, and prevention of damage and stressors on Earth's land, air, and water. Even technology already available today could reduce energy usage in the U.S. by 12 to 22 percent, according to The Information Technology Industry Council (ITI). In the face of this dire reality, the potential of technology to help meet this challenge is a rare source of optimism. According to a recent survey by Intel and the research firm Concentrix, 74 percent of business-decision makers working in environmental sustainability agree artificial intelligence (AI) will help solve long-standing environmental challenges; 64 percent agree the Internet of Things (IoT) will help solve these challenges. As the field of AI develops, so will the potential to protect the environment. From the land and air to both drinking and ocean water, AI is shaping up to be the key that governments, organizations, and individuals can tap to work toward a cleaner planet, even AI can help offices to avoid to waste energy when staffs are working in offices every day.

Many AI scientists indicate that AI will also make renewable energy technology like solar panels and wind turbines more efficient and cost effective, helping them to become ubiquitous and lower society's dependence on fossil fuels. AI will also make renewable energy technology like solar panels and wind turbines more efficient and cost effective, helping them to become ubiquitous and lower society's dependence on the fossil fuels polluting the air — then hopefully eliminate them all together. Combined with the smart grid, another technology that will be enabled by AI, this will truly progress the way people receive and use electricity in their homes, offices, and everywhere else. Smart meters save energy by allowing for two-way communication between the grid and anything that uses electricity, giving energy providers a better understanding of usage and the ability to make real-time adjustments for efficiency. Customers will benefit from the real-time data too; seeing the increased costs at peak times will encourage them to voluntarily adjust their usage to save money. This will, in turn, save even more energy: a win-win. Plus, the process of delivering the energy itself will also be improved by the smart grid, thanks to Volt/VAR control systems that can reduce the amount of energy wasted when it's in electricity transmission lines.

Can AI replace office workers

Can AI replace all office workers to do their different tasks in office different department ? If AI can only replace some department office workers to do their simple tasks, how it can raise more efficiency to compare them in some business office environments. I shall indicate some office tasks to explain how AI can help these businesses to raise their efficiency in officesas below:

- AI insurance workers

Nowadays, some country offices begin apply robotics to replace human office workers in their companies. For example, Japanese company replaces office workers with artificial intelligence in insurance industry. A future in which human workers are replaced by machines is about to become a reality at an insurance firm in Japan, where more than 30 employees are being laid off and replaced with an artificial intelligence system that can calculate payouts to policyholders.

Fukoku Mutual Life Insurance believes it will increase productivity by 30% and see a return on its investment in less than two years. The firm said it would save about 140m yen (£1m) a year after the 200m yen (£1.4m) AI system is installed this month. Maintaining it will cost about 15m yen (£100k) a year. The move is unlikely to be welcomed, however, by 34 employees who will be made redundant by the end of March.

The system is based on IBM's Watson Explorer, which, according to the tech firm, possesses "cognitive technology that can think like a human", enabling it to "analyse and interpret all of your data, including unstructured text, images, audio and video".The technology will be able to read tens of thousands of medical certificates and factor in the length of hospital stays, medical histories and any surgical procedures before calculating payouts, according to the Mainichi Shimbun.

While the use of AI will drastically reduce the time needed to calculate Fukoku Mutual's payouts – which reportedly totalled 132,000 during the current financial year – the sums will not be paid until they have been approved by a member of staff, the newspaper said.

Japan's shrinking, ageing population, coupled with its prowess in robot technology, makes it a prime testing ground for AI. According to a 2015 report by the Nomura Research Institute, nearly half of all jobs in Japan could be performed by robots by 2035. For example, one Japan insurance company, Dai-Ichi Life Insurance has already introduced a Watson-based system to assess payments - although it has not cut staff numbers - and Japan Post Insurance is interested in introducing a similar setup, the Mainichi said. AI could soon be playing a role in the country's politics. Next month, the economy, trade and industry ministry will introduce AI on a trial basis to help civil servants draft answers for ministers during cabinet meetings and parliamentary sessions. The ministry hopes AI will help reduce the punishingly long hours bureaucrats spend preparing written answers for ministers.

- AI public service workers

The automated city: do we still need humans to run public services? If the experiment is a success, it could be adopted by other government agencies, according the Jiji news agency. If, for example a question is asked about energy-saving policies, the AI system will provide civil servants with the relevant data and a list of pertinent debating points based on past answers to similar questions.

The march of Japan's AI robots hasn't been entirely glitch-free, however. At the end of last year a team of researchers abandoned an attempt to develop a robot intelligent enough to pass the entrance exam for the prestigious Tokyo University. "AI is not good at answering the type of questions that require an ability to grasp meanings across a broad spectrum," Noriko Arai, a professor at the National Institute of Informatics, told Kyodo news agency. Hence, AI will have possible to replace some public service workers' tasks.

- AI replace warehouse workers

Denso's use of Drishti shows how some jobs will be transformed by artificial intelligence even when they're unlikely to be eliminated by AI anytime soon. Many jobs in manufacturing require dexterity and resourcefulness, for example, in ways that robots and software still can't match. But advances in AI and sensors are providing new ways to digitize manual labor. That gives managers new insights—and potentially leverage—on workers. For example,some workers say the results are unpleasant. Last year, Amazon warehouse employees in Minnesota staged a walkout to protest how the company uses inventory and worker-tracking technology. They allege that Amazon uses it to enforce a punishing working pace that causes injuries. The company has disputed those claims, saying it coaches employees on how to safely meet quotas.

Workers at Denso were initially wary of the prospect of being video-recorded all day to feed machine-learning algorithms, but Huffman says they have since come to appreciate Drishti's technology. After something goes wrong, workers can now look at the data and video with their managers, instead of having to hope bosses take their account of what happened seriously. Huffman says having a constant readout on productivity also helps managers be more responsive to nascent problems. "If somebody's struggling, not every associate is going to call for help," he says. "If we see their cycle time is jumping through the roof, we can go over and say 'Are you having any issues?'" Workers on Denso lines equipped with

Drishti's technology now get a personal feed of their own data. Monitors on each workstation display how a worker is doing, says Raja Shembekar, a Denso vice president. If the worker completes their assembly step on time, they see a smiley face—if not, a frowny one. Hence, Amazon had begun to apply AI robotic to replace some warehouse workers' tasks.

For another factory manufacture working environment example, AI can replace many manufacture workers to do their tasks in factories. Route 9 skims by Boston and cuts clear across Massachusetts to Pittsfield, a city of roughly 50,000, the largest in Berkshire County. Well east of Pittsfield, Route 9 becomes Worcester Road, named for a city that in earlier times was the nation's largest manufacturer of wire—barbed wire, electrical wire, telephone wire and the wire used in the making of undergarments by the Royal Worcester Corset Co., once the largest employer of women in the United States. Older Worcester residents can still recall the factory bells pealing to signal the start and end of the workday. Now, the bells are silent, and the wire and corset factories have been replaced with three of the nation's largest employers: Walmart, Target and Home Depot. If this sounds familiar, it should. It has been nearly two decades since retail overtook manufacturing as the nation's most important job creator, employing roughly one of every 10 American workers—more people than in health care and construction combined. That's a lot of jobs.

Of course, not all retail jobs qualify as what most of us consider good jobs. Today, the average hourly wage for a nonsupervisory retail worker is $11.24, and less than half of retail workers receive benefits of any kind. Still, as a nation, we've come to a sort of uneasy peace with this trend. We know that manufacturing employs far fewer Americans today than it once did—that iPads and Macs aren't made in America and neither are many televisions, appliances, tools, toys or clothes. We also know that shopping for these appliances, tools, toys and clothes is an all-American pastime: On average, we spend nearly 45 minutes a day (more than 270 hours per year) purchasing goods and services. Retail has become the world as we know it, and many of us expect to make our living working in that world.Thanks to automation and a killer business model, Amazon is so efficient that it reaps nearly twice the revenue per employee of Walmart, despite the fact that Walmart, too, has a substantial online presence. Worldwide, Amazon has installed over 100,000 robots to labor in "perfect symbiosis" with humans in its warehouses and has plans to install many thousands more. While it's not clear what constitutes perfect symbiosis, the robots are said to save the company $22 million annually, per warehouse. The company's master plan of an autonomous future also includes goods delivered by drones and self-driving vehicles.

For while Amazon continues to open warehouses around the globe and staff them with many thousands of human beings, estimates are that every human on the Amazon payroll—whether full- or part-time—displaces two humans at traditional brick-and-mortar operations. And that's a feature, not a bug: As Tim Lindner, a veteran IT analyst, confided in a note to industry insiders, eradicating jobs is the explicit goal of any online retailer. As he once wrote: "Labor is the highest-cost factor in warehouse operations. It is no secret that Amazon is moving to highly automated operations within its distribution centers, and...it has additional technology that can further reduce the number of humans it needs to process customer orders.... You have heard the old programmer's phrase, 'Garbage in, garbage out.'... [With] the diminishing reading abilities of humans on the Receiving dock, finding an automated solution to eliminate the 'garbage in' problem is the holy grail. Amazon may have just patented it."

By garbage, Lindner meant human error, the alternative to which is apparently robotic precision. And robots can be very precise, especially when it comes to routine tasks. Sawyer, an industrial robot created by the former Boston-based Rethink Robotics, offers an impressive illustration of how all-embracing a robot arm can be. Sawyer is the brainchild of Rodney Brooks, the inventor of both Roomba, the robotic vacuum, and PackBot, the robot used to clear bunkers in Iraq and Afghanistan and at the World Trade Center after 9/11. Unlike Roomba and PackBot, Sawyer looks almost human—it has an animated flat-screen face and wheels where its legs should be. Simply grabbing and adjusting its monkey-like arm and guiding it through a series of motions "teaches" Sawyer whatever repeatable procedure one needs it to get done. The robot can sense and manipulate objects almost as quickly and as fluidly as a human and demands very little in return: While traditional industrial robots require costly engineers and programmers to write and debug their code, a high school dropout can learn to program Sawyer in less than five minutes. Brooks once estimated that, all told, Sawyer (and his older brother, the two-armed Baxter robot) would work for a "wage" equivalent of less than $4 an hour.

Robots loom large in discussions of work and its future, a conversation that can get mired in false assumptions. Until recently, many economists were skeptical that automation could permanently displace human workers on a large scale. People have always shifted away from work better done by machines, but the economic principle of "comparative advantage" predicts that humans will maintain an edge in many fields. Under this logic, technology will not displace us but set us free to do less dangerous, more challenging things, essentially the very things that make humans human. Of course, human workers are complicated. We get tired, hungry, distracted, angry, confused. We make mistakes, sometimes egregious ones. Machines lack our frailties and biases and are better equipped to weigh evidence fairly, without prejudice or false assumptions. Perhaps most critically, machines can retain and process data far more accurately than we can, and that data is growing exponentially.

Every minute of every day, Google services 3.6 million searches in the United States alone. Spammers send 100 million emails. Snapchatters send 527,000 photos, and the Weather Channel broadcasts 18 million forecasts. This and more data—properly collected, codified and analyzed—can be applied to automate almost any high-order task. Data can also serve as a surrogate for human experience and intuition. Online shopping and social media sites "learn" our preferences and use that information to make values-based assessments to influence our decisions and behavior. And, increasingly, machines excel in the tasks once thought uniquely human."Computers are able to see and hear, and have face-recognition capabilities that are significantly better than humans," says Vardi. "Machines understand the human world far better than they did just a few years ago. And we haven't discovered anything in the human brain that can't be modeled."

● AI can replace counter cashier service staffs

And robots need not be perfect, only equal to—or a tad better than—complicated and expensive humans. And technologists are working hard to make sure they are a tad better. For example, in the case of retail, it's become clear that many of us avoid the self-service checkout line—we prefer the cashier to punch in our purchases rather than do so ourselves. So it seems that the job of cashier—among the largest retail employment categories—is not directly at risk. But Zeynep Ton, an MIT management expert who focuses on the retail sector, says self-service checkout is only a first step and not a terribly smart one. "Customers recognized that self-service checkout is not an innovation, but merely a way of outsourcing the job to them, so they didn't like it," she says. "But new technology is coming that will make self-service checkout so much easier and faster, and that will have a real impact on retail employment."

Experts caution that the so-called apocalypse in retail predicted a few years ago has not yet come to pass. In fact, for every company closing existing stores, two more are opening new stores. Retail is a highly competitive industry, and technology is transforming not only the way we shop but the way we connect with brands—for example, just a few years ago, who would have imagined that Amazon would open actual retail stores? And while e-commerce has grown to 10 percent of retail, that still leaves 90 percent for brick-and-mortar stores. But those brick-and-mortar stores, too, are undergoing radical change that has serious implications for America's workforce.

As example, Lobaugh cites food trucks, which he says increasingly pose a threat to many fast-food outlets. Unlike restaurants pinned down by a pair of Golden Arches, food trucks are nimble—they can home in on areas where customers are most likely to gather at any particular time. They can also tailor their offerings to a particular region or even a neighborhood, as well as use Facebook or other media to get out the word on their menu items and locations. Small, specialty stores also have far more flexibility than large department stores. "Technology has reduced the cost of entry into new markets, so in retail there are fewer big, monolithic companies, but more small competitors," he says. "Companies are diversifying to meet the specific needs and desires of consumers—everyone's piece is getting smaller, but there are many more pieces."

But despite what it predicts will be a banner holiday season, this year Amazon took on far fewer seasonal employees than usual—100,000 employees versus 120,000 the previous two years. And while an Amazon spokeswoman insisted that automation is not a factor in this reduced workforce, others seem to not agree. In a recent report, Morgan Stanley analyst Brian Nowak soothed the fears of Amazon shareholders concerned with the wage increase by pointing out that automation had already and would continue to reduce the call for labor, and therefore reduce overall costs. When asked about this, Lobaugh again tactfully declined to comment—other than to say that while the retail sector had lost less ground than most people assume, retail employees were another matter. "There are winners," he says, "and then there are losers."

● AI can replace accountants in accountancy service industry

Not that long ago artificial intelligence (AI), robots and machine learning (ML) were thought to be things only found in science fiction films. Today, this type of technology is taking center stage in workplaces across the globe. Industries, including manufacturing, retail, agriculture, and customer service have already had AI replace some job positions that left workers scrambling to find new career options. This AI revolution is not expected to slow down anytime soon. In fact, experts anticipate that as many as 800 million jobs could be replaced with AI technology by the year 2030. Initially, AI technology and automation in the workplace seemed to only affect pink and blue-collar workers. As this technology advances and becomes more powerful, professional, white-collar workers, including accountants, are starting to worry about what the future holds for their career and if AI will be developed to own their professional skills in accounting service industry.

In basic terms, AI technology is intelligent machines that are able to complete repetitive, mundane tasks at a fraction of the time it takes humans and with greater accuracy. The emergence of Machine Learning now allows AI platforms to observe, analyze and self-learn data and processes to improve its performance and accuracy over time. AI technology is already able to handle many accounting functions, such as tax preparation, payroll, and audits. Many of the leading accounting software providers, including Xero, Intuit and Sage have incorporated AI technology into their software to handle basic accounting tasks, such as bank reconciliations, invoice categorization, risk assessment, and audit processes, like expense submissions and invoice payments. Many of these standard tasks are extremely time-consuming, which has many accountants across the country worried about how the emerging AI technology will affect their billable hours. An even bigger concern is that AI technologies will replace the need for companies to work with accountants at all.

- AI Will Transform not Replace Accountants

While there is no doubt that AI technology is capable of handling many standard accounting tasks faster and more efficiently or that these capabilities will only increase over time, it doesn't mean the end for accountants. There always will be a need for that human element - human intelligence - at the other end of AI technology. In fact, according to leading research firm, Gartner, AI is set to create more jobs than it will replace, leaving workers, including accountants with options. Accountants don't have to worry about their job being replaced by AI any time in the near future. Companies will always need accountants that can analyze and interpret AI data, as well as provide consulting services. Rather than replacing the role of an accountant, AI technology will transform the duties an accountant performs.

With AI technology and machine learning handling many of the mundane, repetitive tasks, accountants will have more time to focus on other aspects of the job, such as consulting and data analysis. This is good news for many accountants. Rather than spending hours completing menial tasks, accountants of the future will be able to use and analyze AI data to provide their clients with sound business solutions.

In many ways, AI will help accountants improve their services. AI technology will improve data entry accuracy and lower the liability risk for accountants. In addition, emerging technology is more efficient at fraud detection, adding an extra layer of protection for accountants and their clients. It also provides real-time data, which allows accountants to provide real-time solutions. Even more impressive is the ability of machine learning to analyze large amounts of data instantly, evaluate past successes and failures in an effort to accurately predict future outcomes.

There is no way to escape the use of AI technology, at least not if you hope to remain competitive in the upcoming years. The speed, efficiency and accuracy of AI technology just cannot be beat. The only thing accountants can do is to embrace this new technology and learn how to maximize its use. The better equipped you are to help your clients integrate and utilize AI technology in their accounting processes the more valuable you will be. For example, many universities today are already incorporating IT and database management courses into their accounting program. This means that graduating students are coming into the workforce with the skills they need for future accounting work. Accountants already in the workforce must find ways to acquire these skills in order to remain relevant to their employers and/or their clients. Accountants can obtain the IT skills they need by attending seminars, using self-learning online programs or attending college-level courses. It is equally important for accountants to stay up-to-date on the latest accounting trends, emerging technologies and industry news. This will allows accountants to not only keep their jobs but to also provide more efficient services to their clients. Rather than worry about AI taking over their jobs, accountants should embrace this technology as a powerful solution to enhance customer services. Finally, accountants will be able to use all their training and experience to provide customer will real and effective business solutions, whether it's in reference to tax consulting, real estate deals, mergers, growth options, or any other business practice.

On conclusion, technology is advancing at record rates so now is the time to obtain the IT and database management skills you need to advance into the future. With the right skills and training, accountants are guaranteed a lucrative career that will last well into the future.

(AI) -driven automation industry development

5.1 (AI) - driven automation industry development how to influence work nature change

(AI) -driven automation industry will create wealth and expand economy growth to any countries, but it will be accompanied by changed in the skills that workers need to learn. One of main ways that technology increases productivity is by decreasing the number of labor hours needed to create a unit of output. It implies (AI) technology will influence low educated and low skillful labor number to be decreased ( reduction employment number).

In contrast, technological change tended to work in a different direction throughout the nowadays. The advance of computer and the internet raised the relative productivity of higher skilled workers. So, routine-intensive occupations that focused on predictable tasks disappearance, such as switch board, operators, filming checkers, travel agents and assembling line workers etc. were particularly replaced by new technologies.

However, today, it may be challenging to predict exactly which jobs will be most immediately affected by (AI) driven-automation. The reason is because (AI) is not a single technology, but rather a collection of technologies that are felt unevenly through the economy to influence job changing both negatively and positively. In positively view point, (AI) driven-automation will make many workers more productive and increase demand for certain skills. Consequently, new jobs are likely to be directly create in areas , such as the development and supervision of (AI) as well as indirectly created in a range of areas throughout the economy as higher incomes lead to expanded demand. Otherwise, in negatively view point, many traditional human needed ( demand) skillful jobs will be threatened by automation are highly concentrated among lower-paid, lower-skilled and less -educated workers. It means automation will cause pressure on demand for this group, pressure and employment, if (AI) can replace the low skilled and less educated workers‘ jobs. Thus, (AI) will have negative influence to impact on the labor market.

(AI) capabilities will enable automation of some tasks that have long required human labor. Why can (AI) replace some simple human jobs? For example, advances in robotics are expanding machines' abilities to interact with and sharp the physical world. Combined , (AI) and robotics will give rise to smarter machines that can perform more sophisticated functions than ever before and brings more advantages that humans have exercised. This will permit automation of many tasks now performed by human workers and could change the shape of the labor market and human activity.

5.2 How (AI) influences labor market

Today, it may be challenging to predict exactly which jobs will be most immediately affected by (AI)-driven automation. Because (AI) is not a single technology, but rather a collection of technologies that are applied to specific tasks.

Some specific predictions are possible based on the current (AI) technology. For example, driving jobs and house cleaning jobs, bank counter service jobs, telephone enquiry service operators. Restaurant cooking jobs, simple accounting record service jobs etc. that require relatively less education to perform. Advancements in computer vision and related technologies have made the feasibility of fully appear more likely, potentially displacing some workers in driving-dominant professions. Seemingly similar robot, for which the operational tasks is less specific of navigating to a specific destination when following a set of given rules and preserving safety.

In the future, the effects of (AI) on the labor market in the decade ahead will continue the trend toward skill-biased change that computerization and communication innovations have driven in recent decades. Thus, some human driving occupation will be disappeared or replaced by (AI) automation driven. For example, bus drivers, light truck or delivery services drivers, heavy and tractor-trailer truck drivers, school drivers, tax drivers, travel bus drivers.

However, (AI) technology could enable some workers to focus time on other job responsibilities, boosting their productivity, and actually raised wage growth among those still holding the reshaped jobs. For example, salespeople, who currently spend a considerable amount of time driving could find themselves able to do other work when a car drives them from place to place, or inspectors and appraisers could fill out paperwork, when their car drives itself. This (AI) -driven technology should make these

workers more productive, with (AI) -driven technology serving as a complement, not a substitute. New jobs will also likely be created, both in existing occupations cheaper transportation costs with lower prices and increase demand for products and all the related occupations, such as service and fulfillment, and in new occupations not currently foreseeable.
What kind of jobs will be created by (AI) technology? Predicting future job growth is extremely difficult, due to it depends on technologies or substitute for existing today as well as they may complement or substitute for existing human skills and jobs. However, (AI) will also lead to substantial indirect job creation to the degree it raises productivity and wages, it may also lead to higher consumption that would support additional jobs from high-end draft production to restaurant and retail. The future(AI) " augmented intelligence", the technology's role is as assisting and expanding the productivity of individuals rather than replacing human work. Thus, based on the biased-technical change framework, demand for labor will likely increase the most in the areas where humans complement (AI) automation technologies. For example, (AI) technology , such as IBM's Watson may improve early detection of some cancers or other illnesses, but a human healthcare professional is needed to work with patients to understand and translate patients‘ symptoms, inform patients of treatment options, and guide patients through treatment plans. Shipping companies may also partner workers who pick up and deliver products over the last feet with (AI) enabled autonomous vehicles that move workers efficiently from site to site. In such cases, (AI) augments what a human is able to do and allows individuals to either be move effective in their specially task or to operate on a larger scale. Thus, it seems (AI) technology will also create new jobs, raise productivities and workers' efficiencies.

Redefining management in
the workforce of artificial intelligence

● Change management

In the future, due to artificial intelligence influences to some kind of human jobs nature. So, the kind of human jobs of management methods will also need to change to adapt the artificial intelligence technology input to their organizations. It will cause challenges for every executive and manager if who won't have effort to manage their teams how to apply artificial intelligence technology to work efficiently and easily. For example, division of labor will change among humans and machines will increase. Thus, companies will have to adapt their training performance and talent strategies how to emphasize on work that how to make human judgment and skills and experimentation. Thus, (IA)'s greatest impact will be on administrative coordination and control tasks, such as scheduling , resource allocation.

In fact, mangers will encounter this challenges: How to apply human experience and expertise to judge critical business decisions and practices when the information available is insufficient to suggest a successful course of action? Due to this kind of work will require new skills and mindsets. I shall indicate these change management methods to adapt (AI) technology. Such as: administration and routine tasks, scheduling , allocation of resources and reporting will fall within the intelligence machines, responsibilities that have long been reserved for humans. For example, a typical store manager or a lead nurse at a nursing home most constantly arrange shift schedules, accounting for staff members' absences owing to illness, vacation time or sudden departures.

Thus, the managers need to learn how to arrange new division of labor within the organizations after (AI) technology had been implemented to the organization. Artificial intelligence is currently influencing into once considered exclusive to humans: assessing and acting on human emotions and personality traits. The influences to managers need to change their strategies to adapt (AI) technology implements include such as below:

Firstly, managers need to spend the bulk of their time on coordination and control tasks from intelligent system implements. Their time spending on these major three aspects from impact of intelligent system: coordinate and control, solve problems and collaborate and people and community , strategy and innovation three aspects. Thus (AI) will influence managers need to change their judgment method to teach whose teams how to adapt the (AI) system operations in any organizations.

Secondly, (AI) will influence top, middle and low level management needs to change to adapt the (AI) technology operations to any owned (AI) technology organizations in the future. Intelligent machines must be trained in context. Just like humans , on-the-job training is a requirement for such machines because they typically arrive with only very general capabilities. To get the most from (AI), managers at all levels must participate in the instructional experience and in the learning process and provides managers' familiarity with such systems on these aspects, e.g. How the system works and generate advice, how the system has a proven track record , how the system provides convincing explanations , how the system can make simple rule- based decisions.

Thirdly, managers need to learn how to make judgment more accurate (AI) systems assistance. Although (AI) will invariably take on more routine work and even augment human decision-making, it won't judgment work, the application of human experience and expertise to critical business decisions when the information available is insufficient to suggest a successful course of action or reliable enough to suggest an obvious course of action. For a sense of the nature of judgment work, consider big data marketing and sales analytics. Such analytics often provide insights that can inform promotional campaigns, including predicting which promotions will generate desired sales brand further into the future, marketing executives need use judgment, combining analytics with their own and others' insight and experience.

The application of experience and expertise to critical business decisions and practice represents the real value of human judgment. But, when artificial intelligent machines are implemented to any organizations to assist the low, middle and top level management to make any business judgment. These forms of judgment work that managers can gather data interpretation, idea development more absolute from (AI) machine assistance. Thus, why these level management executives need to learn how to apply (AI) machines to help them to make any business judgment more accurate.

● How (AI) influences organizational change

Consequently creative and social intelligence will be in even greater demand as (AI) makes in management and the workforce. This development will represent a long term trend in labor markets , one characterized by intensifying demand and reward for social skills with a growing desire for creative capabilities, managers will seek to fashion of ideas and hypotheses from inside and outside of the enterprise to shape solutions to their most pressing business problems. Thus, (AI) will influence overall organizational team members who have chance to participate any decision to make more accurate business judgment.

Many managers mistakenly view judgment work as only an individual discipline, failing to appreciate that it can also involve decide interpersonal and organizational practices. In more complex settings, judgment is typically a collective outcome of individuals' and teams' diverse perspectives, insights and experiences. And often , the resulting choices are better informed than decisions that an individual would have arrived at on his or her own.

Thus, when any organizations apply (AI) technology to assist managers to gather data and ideas to make any judgment. In these cases, organizations can create the conditions for effective collective judgment by establishing structures , such as " shadow advisory boards" that prompt managers and employees to source and synthesize multiple perspectives. Thus, a traditional organization (firm) might freshen its thinking is t put together a shadow advisory board, comprised of young, digital people who can apply (AI) machine assistance to make judgment work more accurate whether related to people development, problem-solving or strategizing and innovating for considerable degrees of creative and social intelligence.

Thus, on the one hand, (AI) technology machine augmentation and automation can give these advantages to human (organization managers) , e.g. developing people and community, solving problems and collaborating, coordinating and controlling work, shaping strategy and leading innovation. Besides, on the other hand, the next generation managers need have these individual attitude to treat intelligent machines to be as colleagues.

When, judgment is a human skill, intelligent machines can accelerate human learning that supports it, assisting in data -driven simulations, scenarios and search and discovery activities. Focuses on judgment work, some decisions require insight beyond what data can tell them. This is the sweet sport for human judgment, the application of experience and expertise to critical business decisions and practices. Thus, managers will also need to find ways to learn how to use digital (AI) technologies to tap into the knowledge and judgment of partners, customer external stakeholders and role models in other industries after the (AI) machine had been implemented to the organization.

5.3 Future works change: Automation, employment and productivity

- How (AI) influences employment

Human future " micro to macro" industry trends will be affected business strategy and public policy by (AI) technology. In the future (AI) technology will influence those six themes: productivity and growth, natural resources, labor markets, the evolution of global financial markets, the economic impact of technology and innovation and urbanization. However, (AI) technology will bring economic benefits of tackling gender inequality, a new global competition, Chinese innovation and digital globalization.

Nowadays, advances in robotics artificial intelligence, and machine learning are in a new age of automation, as machines match or outperform human performance in a development to any countries. For example, automation of activities can enable businesses to improve performance by reducing errors and improving quality and speed, and in some cases achieving outcomes that go beyond human capabilities. For example, some research indicated automation could raise productivity growth globally by 0.8 to 1.4 % annually; more than 2,000 work activities across 800 occupations. When less than 5% of all occupations can be automated using demonstrated technologies about 60% of all occupations have at least 30% of constituent activities that could be automated. Many occupations will change that will be automated away: Activities most susceptible to automation involve physical activities, in highly structured and predictable environments, as well as the collection and processing of data. They are most prevalent in manufacturing , accommodation and food service and retail trade and include some middle-skill jobs. For example, such as natural language processing is a key factor. Beyond technical feasibility, the cost of technology competition with labor including skills and supply and demand dynamics, performance benefits including and beyond labor cost savings, and social and regulatory acceptance will be affected by (AI) automation technology. Thus, (AI) automation will impact to influence global employment in those aspects as below:

Firstly, assuming that people are displaced by automation will find other employment. The anticipated shift in the activities in the labor force is of a similar order as the long-term shift away from agriculture and decreases in manufacturing share of employment. Both of manufacturing and agriculture industries which would be accompanied by the creation of new types of work not foreseen at the time.

Secondly, for business, the performance benefits of automation are relatively clear. Thus, the businessmen have opportunities for their micro economies to benefits from the productivity growth potential and macro economies to benefit to encourage continued progress and innovation , investment and market incentives. At the same time, employers must innovate policies to help workers and institutions adapt to the impact on employment.

This will likely include rethinking education and training, income support and safety nets , as well as support for those dislocated, when employees need to leave themselves homes to move to other cities to learn new (AI) automation works. Thus, individuals in the workplace will need to engage move comprehensively with machines as part of their everyday activities, and acquire new skills that will be in demand in the new automation age. Consequently , the scale of shifts in the labor force over many decades that automation technologies can be a similar order to the long -term technology -enables shifts in the developed countries' workforces away from agriculture in the 21 th century. Those shifts did not result in long-term mass unemployment because they were accompanied by the creation of new types of work not foreseen at the time. However, human will still be needed in the workforce when the total productivity gains are caused by (AI) technology.

- What occupations will be influenced by (AI) technology.

In the future, scientists predict that these occupations will be influenced by (AI) technology mostly. They include : retail salespeople, food and beverage service workers, language or translation teachers, health practitioners. Since these work activities have a more relevant occupations are made up of a range of activities with different potential for (AI) automation . For example, a retail salesperson will spend more time interacting with customers, stocking shelves , or ringing up sales. Each of these activities is distinct and requires different capabilities to perform successfully.

Thus, these job activities have similar simple control characteristics. Simple activities include greet customers, answer questions about products and services, clean and maintain work areas, demonstrate product feature process sales and transactions. All these activities can have similar simple activities in order to (AI) machines can be learn how to do these activities from (AI) technology . For example, the capability perception includes sensory perception, cognitive capabilities, such as retrieving automation, recognizing known patterns( supervised learning), logical reasoning problem solving.

Thus, (AI) machine is such human, which has feeling and emotion, such as social and emotional sensing, judgement reasoning methods, natural language understanding and physical capabilities, such as mobility , navigation, gross motor skill, fine motor skills. It seems that the future, (AI) human invents machines which will have these human characteristics to do human similar behavioral job duties more easily and efficiently. It

implies these above human occupations will be replaced by (AI) human invention machines in the future. Due to (AI) creation, it is possible to cause unemployment number of these above workers will increase because (AI) machines can do their similar job behavioral activities.
Consequently, employers won't need to employ many of these skillful labor. Otherwise, they can buy less number (AI) machines to attempt to do whose job activities more easily and efficiently. So, it seems (AI) machines will have more high work performance to replace these occupation workers' work performance. Finally, these occupation worker unemployment number will only increase when the (AI) machines had been invented to achieve to do their work behavioral activities absolutely success in the future.

● Whether (A) technology machine labor will replace human worker more or assist human worker more

There is no single agreed definition of a robot how outcome of a task that is completed without human intervention. When some definitions require the task to be completed by a physical machine moves and respond to its environment, other definitions use the term robot in connection with tasks completed by software , without physical embodiment.
However, to answer the question : Whether (AI) technology machine labor will replace human worker more or assist human worker more. I shall indicate some examples to let readers to judge whether (AI) technology can create new jobs or reduce old jobs.
Firstly, I shall explain what (AI) function is. (AI) is a service robot that performs useful tasks for humans or equipment excluding industrial automation application . Thus, the classification of a robot into industrial robot or service robot is done according to its intended application. It is also a personal service robot or a service robot for personal used for a non commercial task, usually by lay persons . Examples are domestic servant robot, and pet exercising robot. It is also a professional service robot or a service robot for professional used for a commercial task, usually operated by a properly trained operator. Examples, are cleaning robot for public places, delivery robot in offices or hospitals, fire-fighting robot, rehabilitation robot and surgery robot in hospitals. Thus, these functions will be future (AI) application to our daily life necessaries or business

necessaries.

However, some authors agree (AI) will bring negative outcomes of automation, due to raise competiveness, reduce human job nature. Otherwise, other authors argue (AI) will bring positive outcomes of automation, due to raise productivities, job creation, assist humans work.

On the positive outcome hand, robots can increase productivity . This is particularly important for small-to medium sized businesses both are in developed and developing countries economies. It also enables large companies to increase their competitiveness through faster product development and delivery. Increased use of robot is also enabling companies in high cost countries to re shore, or bring back to their domestic base parts of the supply chain that will have previously outsourced to sources of cheaper labor. Currently , the greater threat to employment is not a automation, but an inability to remain competitive. Automation has led overall to an increase in labor demand and positive impact on wages. The reason is that the middle-income/middle-skilled jobs have reduced as a proportion of overall contribution to employment and earnings leading to fears of increasing income inequality, the skills range within the middle income bracket is large. Thus, robots are driving an increase in demand for workers at the higher -skilled and with a positive impact on wages. This issue is how to enable middle-income earners in the lower-income range to unskilled or retain. Finally, the (AI) positive impact supporter who argue the future will be robots and humans can work together.

However, on the negative outcome hand, robots can substitute labor activities, but don't replace jobs. They believe that less than 10% of jobs are fully automatable. Increasingly , robots are used to complement and augment labor activities, the net impact on jobs and the quality of work is positive. Automation can provide the opportunity for humans to focus on higher-skilled, higher-quality and higher-paid tasks. Robots can improve productivity when they are applied to tasks that which perform more efficiently and to a higher and more consistent level of quality than humans. For example, increased productivity is enabling some firms, such as Whirlpool, Caterpillar and Ford Motors company in the US restructure their supply chains, bringing back parts of the manufacturing process to the country of origin. Thus, productivity gains due to robotics and automation are important not just at the company level, but also for build industry and nation competitiveness.

I suppose that productivity can be raised. What are the impacts of robots on employment? Firstly, the main focus of development has been on personal entertainment, which does not drive worker productivity ( manufacturing production). When the internet ( information and communication technology (ICT)) innovation. This is borne and by findings that manufacturing productivity, which has been driven by innovations in automation rather than consumer technologies, has government strongly than productivity in the services sectors of the economy in most nature economies. It seems (AI) automation will create many jobs in internet communication entertainment game industry. For example, many young people like to use internet to play any electronic games from computer or mobile at home or outside home conveniently. Thus, (AI) automation will increase demand to be invented to any new entertainment game from internet channel. It will need to employ many (AI) entertainment game inventors to create many automation entertainment games. Thus, (AI) automation in internet entertainment game industry will need human (AI) entertainment game inventors to invent the knowledge-based capital of (AI) automation entertainment games. The (AI) entertainment game inventors will need own research and development skills, form specific skills, organizational know-how skills, databased knowledge, design and various forms of intellectual property to do these (AI) automation entertainment game invention occupations in the future.

International Federation Of Robotics(2016) indicated that China will be as a major robotics manufacturer and user of robots, benefiting from jobs created by robot manufacturing and productivity gains from robot use. Chins had sold of robots to any one single market every year since 2017 year. The Chinese government has included a focus on robotics in its 10 year strategy. In order to achieve its target of a robot density of 150 units per 10, 000 workers by 2020 year. Thus, Chinese companies will have to install around 650,000 new industrial robots between 2016 to 2020 year, 2.5 times more than installed globally in 2015 year.

Hence, China (AI) manufacturing industry will need to employ many workers . It implies (AI) manufacturing industry will create many new occupations in China. Also, ministry of economy, trade and industry (2015) also showed that Japan currently has the largest stock of industrial robots in operations, primarily in the automation industry. Driven by a rapidly aging population and low productivity rates, the Japanese government has sights on a 20-fold increase in the use of robots in the non-manufacturing sector and a three-fold growth rate of labor productivity in the service sector both by 2020 year. Thus, it also implies Japan will need many robots to be provide to service industry. Due to robots will provide to serve any businessmen's clients. Thus, it is possible that the service workers won't be dismissed as well as it is depended on the serving job nature to decide whether Japan's service workers can still serve to their employer when the service (AI) robots are applied to whose employers.

Consequently, it seems that (AI) can create employment, Ministry of economy, trade and industry (2015) showed that such as China will develop the major (AI) automation manufacturing industry. The (AI) employers will need to employ many workers to manufacture any these different kinds of (AI) robots to satisfy China or overseas individual or business buyers needs. But, (AI) can also cause unemployment to the low skillful service workers. Such as if Japan some service businesses choose to buy any (AI) service robots to replace their service staffs to serve their clients. It is possible that the service staffs will be dismissed, due to (AI) robots can do such as their same service job duties to achieve better service performance. Thus, today, it is increasingly common for people to use robots in various situations at home and in retail stores, hotels and hospitals these service industries. Robots are classified into server types based on their functionality ( service and utility robots or those designed to communicate with humans) and appearance ( humanoid robots or mechanical robots). The type of robot, to which each country allocated particular importance in the advance of robotics, reflects the sense of values and preferences of its population. Thus, if the country has high population needs to use robots, then they will influence either more new jobs creation or more old job loss in the country's (AI) manufacturing or (AI) service industries both. For example, Japan respondents often associate the term " robot " with humanoid robots that can communicate with human and they have a high level of familiarity with robot. The US has the highest level of robot utilization at home and in retail stores with its people being the most enthusiastic about the future use of robots. Germany shows a strong tendency to consider robots for industrial purposes and its people feel strong effort to the presence of robots in their households.

In conclusion, to judge whether how (AI) will influence the country's employment to be better or worse. It will depend on the country home buyers (users) or business buyers (users) how to use (AI) for their daily needs. If the country , such as US retail stores need to use (AI) , it will have possible to reduce some or many retail service workers. Even, if the country

, such as Japan has many home users need to use (AI) , it will not influence the employment market. Otherwise, it will raise (AI) salespeople numbers. Even, if the country, such as Germany and China will have many (AI) manufacturers, then it will create many (AI) manufacturing occupations for these (AI) manufactory workers.

Consequently, (AI) robots manufacturing and service needs will have positive or negative impact to any country's employment. It will depend on the (AI) service provision and service workers' job nature as well as the manufacturing workers of (AI) knowledge level to decide their employment chance in their country's employment market.

5.4 How can (AI) influence labor market?

● How can human society job nature
to be changed to artificial intelligent society?

From the first intelligent perspective reason view point, artificial intelligence is making machines " intelligent" acting as humans expect people to act. Artificial intelligence has ability to distinguish computer responses from human responses, it owns knowledge to solve expert problem. From another research perspective reason view point, artificial intelligence is the study of how to make computers do things which, at the moment, people do better ( Rich & Knight, 1991, p.3).

(AI) researchers are native in a variety of domains, e.g. formal tasks ( mathematics, games), tasks ( perception, robotics, natural language, common sense reasoning), expert tasks ( financial analysis, medical diagnostics, engineering, scientific analysis and other areas).

From the second business perspective reason view point, (AI) is a set of many powerful tools, and methodologies for using those tools to solve business problems. From a programming perspective reason view point, (AI) includes the study of symbolic programming problem solving and search .

From the third human technological perspective reason view point, today's computer can do many well-defined tasks, for example, arithmetic operations, are much faster and more accurate than human beings. However, the computers' interaction with their environment is not very sophisticated yet. How can human test whether a computer has reached the general intelligence level of a human being? Can a computer convince a human interrogator that it is a human? But before thinking of such advanced kinds of machines, human will start developing our own extremely simple " intelligent" machines.

So, it is possible that human society job nature will to be changed to artificial intelligent society when (AI) technology is developed to the mature stage in the future.

- Why does human need artificial intelligence machines?

One of major division in (AI) is between humans who think (AI) is the only serious way of finding out how we ( human) work and human who want companies to do very smart things, independently of how we ( human) work. This is the important distinction between cognitive scientists vs engineers. One of another major division in (AI) is between symbolic (AI), which represents information through symbols and their relationships. Specific Algorithms are used to process these symbols to solve problems or deduce new knowledge and connectionist. So ( AI) , which represents information in network. Biological processes underlying learning, task performance and problem solving are imitated from human mind behaviors.

Thus, it is possible that artificial intelligence machines can do the better judgicious behavior to compare human.

- How does artificial intelligence influence future working changing in automation employment and productivity aspects?

In the automation changing influence aspect, as companies increasingly use robots on production lines or algorithms to optimize their logistics manage inventory, any carry out other core business functions. Technological advances are creating a new automation age in which ever-smarter and more flexible machines will be deployed on an ever larger scale in the marketplace. However, researching artificial intelligence with how influences human working nature. We need to answer these questions: How will automation transform the workplace? What will the implications for employment? And what is likely to be its impact both on productivity in the global economy and on employment?

Advances in robotics, artificial intelligence, and machine learning are growing in a new age of automation as machines match or outperform human performance in a range of work activities, including ones requiring cognitive capabilities. What factors are determined the changing in workplace adoption by artificial intelligence innovation? What advantages are automation? Automation of activities can be enabled businesses to improve performance by reducing errors and improving quality and speed, and achieving outcomes that go beyond human capabilities.

Some scientists indicated based on their scenario modeling. They estimated automation could raise producing growth globally by 0.8 to 1.4 percent annually. Almost, the activities people are paid almost $16 trillion in wages to do in global economy have the potential to be automated by adopting currently demonstrated technology. According to their analysis of more than 2,000 work activities across 800 occupations. When less than 5% of all occupations have of least 30% of activities that could be automated. They also indicated that technical economic and social factors will determine automation. Continued technical progress, for example, in areas such as natural language processing is a key factor beyond technical feasibility , the cost of technology, competition with labor including skills, and supply and demand dynamics, performance benefits including and beyond labor cost savings and social and regulatory acceptance will affect ( alter) the scope of automation.

Other some scientists also indicate U.S. country for example, the anticipate shift in the activities in labor force of a similar order of magnitude as the long term sight away from agriculture and decreases in manufacturing. Share of employment in the United States both which were achieved. So, those factors can influence why artificial intelligence technology needs. So, it is possible that future agriculture and manufacturing both industries will apply (AI) technology manufacturer-kind of job nature to raise productivity instead of farmers, fruit picking workers, farming transportation labours as well as factory manufacturing workers and supervisors etc. human-kind of job nature.

- Is artificial intelligence possible to replace labor ?

Not just intelligence, but also debating, if machines are capable of having a conscious minds. Artificial intelligence has those characteristics as below:

On functionalism aspect, artificial intelligence inputs mental states, sensory inputs, ( beliefs, desires being in pain feeling) and behavioral outputs. Since mental states are identified by a functional role, which are thoughts to be manifested in various systems. Even, perhaps computers which are physical devices with electronic substrate that inform computations on inputs to give outputs similar to brains which are artificial intelligence composed of part any intrinsic relationship to each other. Thus, artificial intelligence activities is not the whole itself, but into parts or on external influence on the parts.

On dualism aspect, artificial intelligence is a set of views about the relationship between mind are matter. On materialism aspect, it builds the only thing that exists is matter, including consciousness.

On biological naturalism aspect, it is similar a human brain than feels pains makes mental situation. So, artificial intelligence is similar biologist which might to be excited to human labor work. Hence, it seems artificial intelligence can change ( alter) or replace human labor work of nature in possible in the future.

- Can (AI) technology replace human labour nature of work?

On technological innovation reason view point, the history development of artificial intelligence studying the intelligence is one of most ancient scientific discipline. The history development of artificial intelligence what aims to achieve human use to sense, learn remember and think, logic probability, decision making and calculation develop from mathematics, instead of replacement human labor functions.

Artificial intelligence history development aim is the scientific analysis of skills in connection and practice with the appearance of computers from 1950 year beginning. The artificial intelligence (AI) can deal with the ultimate challenges. How can ( either biological or electronic) mind sense, understand and manipulate a world that is much simple and more complex than itself? And what if would human like to construct something with such capabilities?

The general-purpose software of the early period of (AI) were only able to solve simple tasks effectively and failed when which should be used in a wider range or an more difficult tasks. One of the sources of difficulty was that early software had very few or mix knowledge about the problems which handled, and activities successes by simply syntactic manipulation. Moreover, the other difficulty was that many problems that were tried to solve by the (AI) were untreatable.

The early (AI) software whether trying step sequences based on the basic facts about the problem that should be solved, experimented with different combinations till which found a solution. From the end the 1960 year, developing the so-called expert systems were emphasized. These systems had ( sue-based) knowledge base about the field which handled. Till to the beginning of the 1970 year, ( Prolog) the logical programming language was born, which was built in the computation realization of a version of the resolution calculus. ( Prolog) is a remarkably prevalent tool in developing expert systems ( on medical, judiciary and other scopes), but natural language parsers were implemented in this language. Then, in 1981 s, the Japanese announced the fifth generation computer system project, a 10 years plan to build an intelligent computer system that use the ( Prolog) language as a machine code. Nowadays, (AI) can be applied any industries, such as car manufacturing industry can use (AI) technological machine-men manufacture car, instead of replacing human labors in factory. Even, in the future, using (AI) machine-men drivers can drive any private cars or public transportation tools, instead of replacing human drivers, e.g. bus, train, tram, ferry etc. Also in the future, machine-men can replace housewives to serve families to do housekeeping clean job , e.g. cleaning toilets, bathrooms, kitchens, even cooking functions at home. So (AI) machine-man can reduce housewives works at home. Moreover, (AI) machine man can take care old people , when who are living at homes or elder care centers.

So, it seems artificial intelligence (AI) will be possible developed to manufacture a new generation machine-man to assist ( serve) families to do any simply cleaning or cooking jobs at homes. Moreover, the overall demand of ( AI) general social needs will also rise, such as security, driving transportation tools, restaurant cleaning, elder centers care service etc. So, it seems that individual or families or social needs of (AI) will be increase in the future. Thus, it will influence macro economy growth (GDP) if there are large house family consumer group and hotel or bus or taxis or ferry etc. different business consumer group demand any artificial intelligence machine numbers increasing. Then, the artificial intelligence products and material manufacturers must need to buy many artificaial intelligence materials to produce any kinds of artificial intelligence machines to prepare to satisfy consumer individual needs. Consequently, macro economy will grow to the owned artificial intelligence development countries, e.g. US, China, UK.

● Why can artificial intelligence satisfy human needs?

First, On machine-man satisfactory demand aspect view point, it makes computers that think, it is the automation of activities. We associate with human thinking: like decision making, learning. It is the act of creating machine that perform function that require intelligence when performed by people. It is the study of mental faculties through the use of computational models. It is the study of computations that make it possible to perceive, reason and act. It is a branch of computer science that is concerned with the automation of intelligent behavior. It is anything in computing service that human don't yet know how to do property.

Second, on thought aspect artificial intelligence means systems thank think like humans, systems that think rationally.

Third, on behavioral aspect, artificial intelligence systems that act like human and that systems act rationally. However, the basic objective of (AI) is to represent human's thought processes in computation . These machines are supposed to exhibit behavior that. It is performed by a human being, would be considered intelligent. However, some authors feel (AI) has disadvantages, such as it is not creative, it is excited in the use of sensory devices, it can't make use of a very wide context of experiences and it does not use common sense.

For speech recognition and understanding function needs example, (AI) can be applied in speech recognition and understanding function, which (AI) speech or voice recognition is a data input method. For example, the computer recognizes and understands one ( or a few) word commands. Speech understanding on the other hand is the computer's ability to understanding a spoken language. That is , the computer understands the meaning of sentences, an paragraphs through (AI).

So, (AI) can be attempted to learn human language how to speak. It is similar to translate human language skill, instead of actual human speaking skill. Also, (AI) can assist handicap learning or language student how to listen different languages by machine-man sounds from computers more accurately. So, it seems that it (AI) can replace human language teachers speaking function and can change teaching language nature of job in language speaking and listening education industry.

- Is artificial intelligence one good choice for human future technological benefit?

Nowadays, new technology development is popular. However, artificial intelligence is one kind of new technology choice among different technologies innovation. So it brings this question: Is artificial intelligence technology value to invest? To answer this question. I shall indicate some other new technology developments to compare (AI) technology development to judge which has urgent needs to achieve human expectation nowadays.

For example, why is green peace interested in new technologies? New technologies features prominently in our ongoing campaigns against genetic modified crops and number power. However, which are also an integral part of our solutions to environmental challenges, including renewable energy technologies, such as solar, wind and wave ( water) power energy as well as waste treatment technologies, such as mechanical, biological treatment.

It seems humans need concern how to apply (AI) technology to solve environment pollution challenges in our future. So, environment protective, agriculture, natural energy technology will be popular demand to attempt to apply (AI) technology to solve their challenges or apply (AI) to assist to develop their industry.

- How can artificial intelligence impact on workplace?

Modern information technologies and the labor economy growth of machines is powered by artificial intelligence have already strongly influenced the world of work in the 21 ST century. Computers, algorithms and software simplify every tasks and it is impossible to image how most of our life could be managed without them. How can be the information economy characterized by exponential growth replaces the most production industry based on economy of scales? What will the future world of work look like and how long will it take to get? Will the future world of work be a world where humans spend less time earning their livelihood? Alternatively, are mass unemployment, mass poverty and social distortions also possible scenario for the future, where robots, artificial intelligence systems play an increasingly central role? These questions concern how artificial intelligence further development . Can influence labor economy growth on workplace ? When the labor market has widespread impact on intelligence property, information technology, product liability, competition and labor and employment laws.

How (AI) technology impacts on labor workplace.

The future influence any organizations how labor economies use of (AI) can be analyzed, such as deep machine learning is based on a set of model high level data. Unlike human workers, the machines are connected the whole time in workplace. If one machine makes a mistake, all autonomous systems will keep this in mind and will avoid the same mistake the next time.

Over the long run intelligent machines will win against every human expert. Production robots have been replacing employees because of the (AI) technology. They work more precisely than humans and cost loss. Creative solutions like 3D printers and the self learning ability of these production robots will replace human workers, the automatic data recording and data processing, traditional back office activities are no longer in demand. Autonomous software will collect necessary information and will send it to the employee who needs it. Additionally, dematerialization leads to the phenomenon that traditional physical products are becoming software. For example, CD or DVDs are being replaced by streaming services. The replacement of traditional event ticket, e-travel ticket service products or hard cash will be the next step, due to the possibility of payment by smartphone. So, (AI) technology will impact human's daily life consumption behaviors in the future. For another example, transportation tools, such as boats and ferries and private vehicles will use sensors and navigating without human input. Taxi and truck drivers will become obsolete, the stock store applies to stock managers and postal carriers of the delivery is distributed by (AI) machine delivery method.

What is the relationship between (AI) and (CRM)?

- Can (AI) technology impact on customer relationship management (CRM) ?

Nowadays , (AI) is a technology almost as old as the computer industry itself, it is similar with the advent of personal assistants function to businesses and personal promotion channel, such as ( Amazon's Alexa, Apple's Siri, Google's Assistant) image recognition ( face book), personalized recommendations ( Netflix , Amazon). Those innovations have been driven by a increase in processing power, lower cost hardware, and the exploding creation and availability of data. It seems, (AI) technology can impact global customer service management method.

How to forecast economic impact modeling to (AI) will affect global economy? Can human forecast business revenue growth and job creation ( or destruction) based on (AI) applied to customer relationship management (CRM) activities? In addition to the economic impact on (AI) or (CRM) which can include an estimate of the economic impact attributable to sales forces customer base. What can economic benefits be brought to (CRM) from (AI) technology?

Artificial intelligence(AI) comprises a set of technologies that use natural language processing, machine learning, knowledge graphs, and other tools to answer questions, discover insights and provide recommendations. Computer systems can use (AI) hypothesize and formulate possible answers based on available evidence can be trained through the ingestion of vast amounts of content, and automatically adapt and learn from (AI) self mistakes and failures.

So, any business organizations (customer service departments) can provide efficient and effective customer relationship management of excellent customer service quality if which applied (AI) technology system. The different type of (AI) systems include: (AI) system platforms, machine learning (AI) based data preparation and enrichment tools, machine vision/ image recognition, voice speech recognition, text analysis and natural language processing, bots , e.g. face book website and virtual digital assistance solutions, social media pattern analysis , sentiment analysis, advanced numerical analysis (e.g. IOT streaming , machine logs), supporting technologies, knowledge base dialog management, Q&A processing etc. different (AI) technology system customer relationship management (CRM) tools.

(AI) (CRM) of activity can include these categories, such as: corporate marketing, marketing operation, field marketing, customer support, digital commerce, customer analytics, customer influenced product or service design, product or service pricing, finance information, presentation, customer billing, inventory , logistics and fulfilment support, partner management etc. different CRM tools.

(AI) technology of CRM has been carrying on plan different stages to achieve CRM personal assistant tool for businesses. The stages are such as, in the beginning stage of (AI) projects in place, implement now, pilot phase next year in the final stage of (AI) customer relationship management tools are foreseeable future. So, this CRM technology has been improved to plan in different stages every year to prepare to achieve full capacity of CRM service quality for businesses to use in the future.

Hence, how to develop an estimate prediction of the economic impact (AI) technologies could have CRM activities, which depends on gathering macroeconomic information on business revenue and the basic marketing of business revenue and the basic markup of business expenses by major functions ( customer support, marketing and sales , production etc.)

An economic impact model that can gather data together and forecast the results how (AI) artificial intelligence technology brings (CRM) customer relationship management benefits to businesses, e.g. surveys investigation includes IT spending by sample countries, GDP and population estimates and forecasts, revenue per employee and ratios of IT spend to GDP. Surveys ( questionnaire questions) of forecast results are influenced by (AI) impact can include: results are projected from surveys and rely on estimates are made by respondents on the expected financial improvements in categories of (AI) –assisted customer relationship management activities. The forecast assumes that these estimates are correct; financial estimates are based on estimates of "first year" improvement from full (AI) implementation;

forecasts are from planning to implement any artificial intelligence of customer relationship management (CRM) projects, the improvement forecast is of categories of activity , e.g. corporate marketing , digital commerce, and customer analytics. They are not estimates of ROI for the (AI) software. They rely on conservative estimates to which each of these entities might affect company revenue, expenses or productivity. They also rely on estimates of the penetration of software in customer relationship management activities . Net new jobs created are based on the ratio of new revenue to jobs required to support that revenue . They can assume that 50% of the net new revenue will support increases in labor and the rest will go for capital and other operating expenses that may replace jobs lost to automation.

In the future, some of the ways in micro economic benefits to any organizations. (AI) technology is expected to impact CRM activities include: Spending up sales cycles, improving lead generation and qualification solving customer support problems faster ( raising service quality), helping companies improve brand campaigns and recognition, lowering costs of support calls when increasing resolution rates, lowering the cost of recruiting employees and partners, increasing revenue from optimized product marketing, optimizing price, distribution logistics and preventing loss through fraud detection. So, micro economic benefits view point, it seems that (AI) CRM technology can raise any companies economic benefits for care term.

Artificial intelligence enables machines or the in-build software to behave like human beings which allows these decisions and act. The advent of (AI) is leading , talking, making decisions and act. The advent of (AI) is leading to new technologies advances and transforming the economic and employment opportunities for humans in a positive way. (AI) related technologies can facilitate our live. For example, industrial robotics, robotic medical assistants, smart games, financial forecasting software, big data analysis, algorithms in health and bioinformatics, pilotless cargo places, drone ambulances and general purpose and workplace robots and others. (Disruptors technologies: Advances that will transform life, business and the global economy).

Artificial intelligence also known as computational intelligence is defined as “ the human –like intelligence exhibited by machines or software. It is theorized that intelligence of humans can be described and intelligence machines or software can simulate it. These machines software can be reasonable , learn, perceive and process information, like human mind and thus facilitate human life. They can think and act for us. So, artificial intelligence is an interdisciplinary field of study including computer science, neuroscience, psychology, linguistics and philosophy.

However, (AI) research and developments have economically impacted many industries, such as robotics, telecommunications, computer applications , health, finance, heavy manufacturing, transportation, aviation, e-service and e-commerce, military , music and movie, toys and games entertainment etc. industries.

In fact, many ideas, systems and technologies have been developing in the world of (AI) technology. However, which are net called or considered (AI) products, rather which are mentioned with their specific names, such as smart graphics, machine learning, e-commerce etc. ( i.e. this is called (AI) effect).

- How does ( AI) technology influence
the future of employment change?

Are future nature of jobs changed to computerization from (AI) technology? Where are the probability of computing occupations from (AI) technology influence? What is expected impacts of future computing on labor market from (AI) technology influence? John Maynard Keynes's frequently cited prediction of widespread technological unemployment " du to our discovery of means of economic the use of labor outrunning the pace of which we can find new used of labor" ( Keynes, 1933, p.3).

In the future, (AI) technology will impact some nature of occupations to change computing. This chance will also influence some countries' economic change. For example, some factory human labors hand routine manufacturing tasks will be changed to computerization of routine manufacturing tasks by (AI) technological machine men hand manufacturing method. it will cause a structured shift in the labor market, with workers reallocating their labor supply from middle-income manufacturing to low-income service occupations.

Arguably, this is because the manual tasks of service occupations are less computerization, as who require a higher degree of flexibility and physical adaptability. So, (AI) technology will influence the human hand labor skillful occupation nature of task cheaper , such as vehicle manufacturing , ship manufacturing, computer manufacturing, steel manufacturing, television, radio etc. home electronic products of heavy machine industry change. Due to (AI) technology machine man will be proper to be used to manufacturing these electronic products when the (AI) technology innovation can develop to the mature stage. Then, any countries manufacturers will choose to use (AI) technology machine man, instead of human hand production.

Supposing the future prices of computing are fallen, seriously, problem solving skills are becoming relatively productive, explaining the substantial employment growth in manufacturing occupations, involving cognitive tasks where skilled labor has a comparative advantage, as well as the increase education needs for (AI) technology computing of machine man subject study.

Prediction of education needs for (AI) technology student numbers will increase, due to manufacturing industry needs many (AI) technology students in future employment market. Another (AI) technology influence if the future (AI) technological innovation, e.g. machine man manufacturing or machine man service industries will both increase demand, then with more sophistic software technologies will be disrupted labor markets by marketing workers redundant.

For publishing industry, what is striking about the case in paper book publishing industry will be unpopular? Due to the electronic book publishing industry will be popular, e.g. Amazon publish . (AI) technology can influence paper book manufacturing method which is replaced by machine man electronic book manufacturing method as well as it will cause the computerization is no longer confined to routine manufacturing tasks. Due to (AI) machine man manufacturing technology will be proper to be used to manufacture any products in short time efficiently and effectively , e.g. electronic book products. In the future, if it is fact to occur this case, such as ( AI) technological machine man manufacturing method will be adopted ( applied) to manufacture electronic books or any products in possible. (AI) technology will cause many manufacturing workers are unemployed. It is beneficial to employers, who can reduce to spend much wages expenditure to employ manufacturing workers, but it will cause many manufacturing workers loss jobs and reduce income to support whose families lives. It will cause social challenges, e.g. increasing stealing crimes if the manufacturing workers had not other skills to find other jobs to do easily. So, manufacturers need to concern over technological unemployment which will be hardly future phenomenon if who decided to dismiss all manufacturing workers, due to (AI) technology machine men replace to them.

If ( AI) technology can be innovated to produce any kinds of machine man to serve any service or manufacturing industries successfully. Then, it will bring these questions: Can future that workers be influenced to be automation employment and productivity by (AI) technology influence? Does it impact to influence the (AI) technology countries‘ productivity and growth and natural resources development and labor markets and evolution of global financial markets and economic impact of technology and innovation and urbanization etc. issues? How will automation transform the workplace? What will be the implication for employment? What is likely to be its impact both on productivity in the global economy and on employment?

In fact, automatic of activities can enable businesses to improve performance by reducing errors chance and improving quality and speed, and same cases achieving outcomes that go beyond human capabilities. Some economists indicate (AI) technology would give a needed boost to economic growth and prosperity have of the working age population in many countries. Based on the scenario modeling, they estimate automation could raise productivity growth globally by 0.8 to 1.4 % annually. They also indicated that almost half the activities people are almost $1.6 trillion in wages to do in the global economy have the potential to be automated adapting current demonstrates technology, according to their analysis of more than 2,000 work activities across 800 occupations. When less than 5% of all occupations can be automated entirely using demonstrated technology, about 60% of all occupations have at least 30% of worker made activities, that would be automated. More occupation will change to be automated. They also indicated for business performance benefits of automation are relatively clear, but the issues are more complicated by policy making to attract foreign investors. Beyond technical feasibility, the cost of technology, competition labor will include skills and supply and demand dynamics, performance benefits and beyond labor cost savings and social and regulatory acceptance will affect the automation. Their predictions suggest that half of today work activities could be automated by 2055 year, but this could happen 10 to 20 years earlier or latter depending on the various factors in addition to their wider economic condition.

Some scientists suggest (AI) technology is finally starting to deliver real-life business benefits. Computer power is growing significantly , algorithms are becoming more sophisticated and perhaps most important of all, the world is generating vast quantities of the fuel that powers (AI) technology data billions of gigabytes of it every day. Also, online firms are digital natives, such as Google online search service company is investing on (AI) technology. For new though most of the news if coming from the suppliers of (AI) technologies. And many new users are only in the experimental phase. Few products are on the market or are likely to arrive these soon to drive immediate and widespread adoption. As a result, analysts believe (AI) technology's potential will give true economic benefit in the future. (AI) industry will introduce to suppliers and users to raise economic potential of (AI) technology.

In the future, (AI) technology systems can solve business problems. Some scientists categorized those into five technology systems that are key areas of (AI) technology development: robotics and autonomous vehicles, computer vision language virtual agents and machine learning , which is based on algorithms that learn from data without replying on rules-based programming in order to draw conclusions or direct an action.

Such as computer vision and language includes natural language processing, analytics, speech recognition technology, some are about learning from information, such as about machine learning and others are related to acting on information, such as robotics, autonomous vehicles and virtual agents, which are computer programs that can converse with humans. Machine learning and a subfield called deep learning are artificial intelligence applications.

● Can artificial intelligence impact
global office productive efficiency ?

Artificial intelligence (AI) is a term first defined in 1956 year. It is a branch of computer science that aims to create intelligent machines that work and react like humans. In contrast today, 60 years later, (AI) is characterized by a number of applications, including computers playing games against humans and understanding human languages, virtual personal assistants, and robotics which involve computers seeing , hearing and reacting to sensory stimuli. In the future, technologists predict for (AI) technology ranging from (AI) being used as a tool to aid relatively simple processes for robots with human like mental capabilities, who expect (AI) technology can emulate human performance by learning, coming to mind its own conclusions, understanding complex content, engaging in dialog with people, enhancing human cognitive performance or replacing humans in executing both routine and non-routine tasks. In existing industry, (AI) technology is used , such as targeted advertising and virtual used personal assistant as well as the (AI) technology that my exist in the future, such as robots with human vehicle processing capabilities.

The range of (AI) technology's progress in the future will determine the economic impact future of (AI) technology on the global economy with more limited advances and applications ( i.e. weak (AI) only) corresponding to more limited economic impacts and more substantial progress, i.e. strong (AI) technology is corresponding to more significant economic impact.

(AI) technology learning that automates analytical model, including predicting cause-and-effect relationship from biological data, identifying new drugs, self-driving cars and protecting against fraud etc. functions. Also (AI) learning can improve natural language processing that allows computers to continue to better analyze, understand and generate language to interface with human using the natural human language, virtual personal assistant, helps users by providing scheduling appointment, reminds organizing personal finance and finding providers of various services, machine vision allows (AI) machine man to identify object, scenes and activities in detect pedestrians and bicyclists.

We expect the economic effects of (AI) technology to include both direct GDP growth from sectors that develop or manufacture (AI) technology and indirect GDP growth through increased productivity in existing sectors that employ some from of (AI) technology. If (AI) producing sectors could grow, then it could lead to increase revenues and employment of (AI) technological professionals within these existing firms as well as the potential creation of entirely new economic activities to any countries' societies productivity improvement in existing sectors could be realized through faster and move efficient processes and decision making as well as increased (AI) technological knowledge and access to information available in societies easily.

In the future, if (AI) technology is an increasingly critical component of more products, it will become an integral part of necessary products of many people's lives. The extent of (AI)'s economy effort is also likely to vary from region to region, thought variation may be more dependent on the predominate economic activity of a region and the (AI) ability can influence economic activity, rather then the economic or developmental status of the regions. (AI) technology can move accessibility and can use source development to do international business between one country and another country.

So (AI) technology has the potential to give benefits to different income chooses and to bring significant gains to both developed and developing countries. For agricultural technology, (AI) has the potential to optimize food production around the world by analyzing agricultural regions and identifying what is necessary to improve crop yield. In total, (AI) technology gives greater economic impact to any countries agricultural regions if which implemented (AI) technology to grow crop , fruit etc. food production in the farms.

Investment in (AI) technology is such as capital investment to any countries‘ public or private enterprises. So, it will have large economic impact to the future . If the (AI) technology is reasonable invested to the different needs aspect by the public or private enterprises in the country. Then, it will have good economic impact to the country in the future. However, when (AI) technology is likely to affect both the productivity and employment components of economic growth in many sectors. Significant public debate has focused on projections of (AI)'s effect on the labor force. However, for instance, some researchers have argued that the rise of (AI) technology and automation will led to significant unemployment as capital is substituted for the low skillful labor. So, they point to the concern that the increasing sophistication of (AI) technology may balance skilled and semi-skilled workers and the reduce the size of the middle class. However, this is not a new argument, due to (AI) technology negatively affecting the labor force and leading to mass unemployment. Because the (AI) technology is the substitution of machinery for human labor. Although, employment in certain industries, has been reduced in the past due to technological advancement. For long term, the labor market has adapted to the introduction of new technology, giving rise to new jobs in new areas. (AI) technology may also be accomplished without a reduction to total employment in the long-term to some Asia countries, such as Hong Kong and Japan. Because Hong Kong and Japan many low skilled labor, e.g. security, cleaner who complaint that employers need them to work long time hours. ( abnormal working hours) e.g. one day 12 to 15 working hour per day. Hence, if (AI) machine means invention technology success. Security or cleaning job can be worked from (AI) machine man in some hours every day in order to reduce the long time working hours cleaners or security workers, e.g. one (AI) machine man works 4 hours for cleaning or security job, one day as well as another cleaner or security labor only needs to work 8 hours one day. So total security or cleaning employers can employ 12 hours machine cleaners or security workers and human cleaners or security workers in one day. For long term benefit, Hong Kong or Japan every security or cleaning worker does not need to work 12 hours minimum working hours one day. They won't feel tried and bore and without private with whose families, so who will accept to do these cleaning or security jobs, even they can raise work efficient and performance when who feel happy and health.

So, (AI) technology of machine man invention can raise low skillful labor efficiency and it can help them to avoid abnormal working hours demand in some busy work life countries, such as Hong Kong and Japan. Before, one Japan female labor feel unhappy to work, due to who often needs to work abnormal working hours for her employer and who has less sleeping and without any private time to enjoy her life with her families every day. So this abnormal working hours factor causes her to do commit suicide behavior, then she is die unlucky. So (AI) technology of machine man invention ought avoid abnormal working hours demand for employer in any countries in the future.

The most important occurrence to any employers, some researchers had attempted to do one experiment to find that private research and development , venture capital and public research and development investment all have strong net effect or economic growth with venture capital funding further having the strongest such effect from (AI) technology. The researchers hypothesize the venture capital investment contributes to economic growth through (AI) technology innovation and by the capacity of an economy to use existing (AI) technology knowledge to increase productivity. They predict the impacts of venture capital, business-research and development and public research and development can raise multi factor productivity from (AI) technology introduction.

Can (AI) technology influence the economic development to developing countries? The developing regions of the world contain most of natural resources. If one day, (AI) technology has invent one kind of machine man which can assist any gas or oil workers to seek any new oil/gas natural resource locations easily. I believe that (AI) technology can help these natural resource exploitation countries will gain economic benefit more easily. So, (AI) driven technology can be used to change to create any new opportunities to address poor management or resources and improve human well being, such as Africa Latin America and India can use (AI) technology machine man to seek any oil/gas natural resource countries exploitation activities to attempt to gain much economic benefits.

● Can AI help offices to reduce labor number

Nowadays, increases in capital and labor are no longer driving the levels of economic growth, such as (AI) technology. The ability of increase in capital investment and in labor of traditional drivers of production, have no longer to be enjoyed in most developed economies ,e.g. developed country, US, UK . However, artificial intelligence has the potential to overcome the physical limitation of capital and labor to avoid missing out on this opportunity. So, policy makers and business leaders must prepare for and work toward a future with artificial intelligence. They must do with the idea that (AI) is another simply method to enhance productivity method . Rather they must see (AI) as the tool that can transform thinking about how growth is created.

Economists have always thought of new technologies are as driving growth their ability to enhancing. It can replace labor and capital factor of production. So, it brings this question: What is the factor of production (AI) technology characteristics. They key factor is to see (AI) technology as a capital-labor .

(AI) can replicate labor activities at much greater scale and speed, and to even perform some tasks began the capabilities of human. For example, by using virtual assistants , 1000 legal documents can be reviewed in a matter of days instead of taking three people six moths to complete. Some (AI) technology may be one kind of factor of production in the future. For another example, people will work in workplace digitalization environment. So, in the future, working environment and information management are automated. Such as Konica camera sale company will use workplace digitalization. So , (AI) technology can provide workplace digitalization in order to raise productivity efficiency. (AI) technology will be one kind of production which is replaced by workplace digitalization and it will grow any organization productivity efficiently. Then, (AI) technology will assist overall social economy growth , due to productivity is raised and products can be produced in short time to prepare to sell in consumption market. So, time will be shortened to increase GDP growth fast for the development of (AI) technology countries.

What will be the development of (AI) technology and predictions concerning the future evolution? The computers and robots will develop conscious, intelligent and minds into humans, enhancing psychological and behavioral abilities and allowing for direct communication with (AI) minds.

(AI) technology will be impacted human life by (AI) technology information communicative and environmental influence. A " world brain" and " world mind", this psychological system will be enhanced and enriched the capacities of both individual and collective cognition by (AI) technology of service industries.

(AI) technology with influence these human needs of service industries changes, such as , biological science, finance, entertainment, business, biological science, transportation, communication military etc. The personal computer evolution, the internet and the world wide web which exploded on the scene, linking business, homes, schools, social organizations which were a completely unpredicted phenomenon to influence human life. Kurzweil (1999) predicts that by 2029 year, most human communication will be with machines. According to Person, by 2100 year, there will be human machine convergence.

How can (AI) technology influence environmental protection to make benefits to farming economic growth? (AI) technology can be applied to predict how to solve environmental pollution challenge to avoid to damage any crop or vegetable or rice or fruit etc. food growth. Because environmental experts can gather global environmental pollution data from an environmental database to build a perform a systematic analysis from (AI) technology. The first step is this broad analysis can include understanding, statistical and data gathering techniques to obtain the relevant data, the correlation among the variables involved, and a list of possible models. The next step is to select a set of methods and models that cover all kinds of knowledge and functionalities needed for the decision making process. Once the models are selected, they must be fully implemented by means of machine learning , data mining, statistical or numerical technique. After that, those models must be integrated to build the whole EDSS. The EDSS must be tested to check its performance, accuracy, usefulness and reliability, both from the user's and (AI) technology/computer scientist's point of view. If these is any wrong feature in any development stage, such as model's integration, models' implementation, selection of models, database, problem analysis etc. the developers must come back in the update th required components. When the evaluation phase is all right, the EDSS is ready to be applied to the environment. The great contribution of artificial intelligence to EDSS the integration of several methods complementing the classical statistical models/simulation , statistical analysis, linear models, etc. and numerical

models ( control algorithms, optimization techniques etc.) .
This cooperation makes the resulting systems more reliable and powerful in coping with real world environment systems. Date interpretation has been a principal area of research in (AI) technology since the very beginning. The most demanding problem in the environmental assessment context. Knowledge representation permits the definition of the different types of data that the existing methods adapt to the process. There is also a lot of work to clean, repair and transform the huge available quantities of raw data. Apart from this, the availability of meta-information or background knowledge is required to guide the process. Data mining is multi-disciplinary: It covers expert systems, data based technology, statistics, data visualization and unsupervised machine learning. These techniques operate at the level of data and background information, where numerous and often incompatible new commensurate pieces of information from disparate sources have to be brought together ( K, Fedra, 1994).
So, it seems that in the future, (AI) technology with the increasing maturity in particular those related to knowledge and engineering, new dimensions can be assisted to users in environmental decision making are available. For example, many environmental systems are characterized both by incomplete models and by limited data. Hence, in the future, (AI) technology will be applied to predict climate change to reduce crop or fruit etc. food agriculture challenge by climate change bad influence.

To understand how the manufacturing business must adapt to prosper in the technology, we need to understand how (AI) technology will change us to shape our daily habits to satisfy our expectation of products to how we shop and even the immediate of the entire process. For example, taxi services are in the crosshairs as on demand transportation services like, available of the touch of a smart phone button expand. In fact, Yellow lab, US country , san Francisco city's largest taxi company is filing for bankruptcy as the industry starts to change faster than almost anyone expected. However, at this point, its more than an app that is changing, some our taxi passengers renting taxi transportation to catch consumption behavior.
(AI) technology will influence digital economy for taxi passenger's individual customer experience, offering a growing renting taxi to catch of service and feedback opportunities when any one taxi passenger who chooses to use mobile phone app online tool to prepaid to rent any taxi more easily.

Also in the long term, (AI) technology can influence vehicles drive themselves of behavior. Already, companies like Google and GM are working on projects to bring fleets of autonomous vehicles to cities at the path of a button.

Moreover, this on-demand service model is beginning to appear across a much broader range of markets. For example , Amazon company is investing in its own fleet of trucks, planes and even drone at the same time as it pushes for same-day delivery of products. As some point, vehicles will be autonomous too. So, it seems that (AI) technique will influence any transportations choose to use digital autonomous driving technology in the future . For Amazon company case, it is not stopping of logistics. It is also aiming to automatically manage the supply of consumer home products with its recently launched Amazon replenishment service, Dash. Dash is a digital service that enables that connected derive to automatically order physical products from Amazon when supplies are running low. So, it seems (AI) technology will be applied to logistic function by digital technology method introduction in the future.

Hence autonomous vehicles will optimize industry supply chains and logistics operations through increased efficiency and flexibility. In fact, fully automated and lean supply chains will keep reduce load sizes and inventory by leveraging smart distribution technologies and smaller autonomous vehicles by machine man assistance. If Amazon continues to grow market share for online sales by reducing effort required by the consumer to place an order, when also contributing the almost immediate delivery of products to the doorstep. So, it will further fuel the trend toward on-demand derive. As Amazon company fuels the on-demand economy, consumers will expect immediacy in more parts of the digital economy. On top of speed, consumers increasing expect more personalization options.

So, (AI) technology will influence digital manufacturing, such as Amazon publishing to monitor every aspect of every process in real -time and communicating to self-optimized deep learning robotics, new methods of high volume and high customization will become possible. Then, as products merge into product platforms and even services, manufacturers have the opportunity to provide components and platforms used by smaller players. So, (AI) technology will influence manufacturing industry to choose automated SMI lines, robots installed, automation engineers.

Another future (AI) technology development can be applied to space science aspect, such as Automation engineering space in manufacturing process to achieve digital manufacturing benefits to any businesses in the future. Such as reducing cost, shortening manufacturing time, raising efficiency, shortening delivery products to client individual time. How can artificial intelligence give the need and advanced fast and evaluation methods benefits for space exploration? When US NASA ( space exploration organization) achieves any space exploration missions, it will answer this question:

When is it useful to have a machine use (AI) technology to achieve a decision? After all, after millions of years of space exploration and rough 10,000 years of civilization, humans are usually quite good at making decisions in complex uncertain environments. Through, Johns Hoplains University's Applied Physical Lab. Research in (AI) technology enabled systems, which has identified three general use cases for (AI) technology to explore space mission:

First, for some tasks (AI) technology is more cost effectiveness than human. Second, (AI) technology is better suited than humans at solving some, but not all problems. Third, (AI) technology allows NASA organization's space exploration mission to develop machines that ate capable of responding faster than when a human is in the decision loop ( D. Scheidt, 2012, A. Castano et. al. 2008).

So, the use of (AI) technology to enable science by observing the pace of rapidly evolving phenomena was demonstrated. It is more effectively coordinating and (AI) technology utilizing to earn economic benefits to use for space exploration mission.

However, (AI) technology also have current risk for space exploration. Today (AI) technology is immature and requires further development to reach its potential. For instance, the (AI) technology algorithms that detected the dust derive could not have identified whether the Martain weather represented a threat to the cover. Also it can not yet use instrument input to determine what, where and how to autonomously make the next space science measurement. An equally important factor limiting (AI)'s deployment is that lacks the methodology and technology to effectively test (AI) technology. So, the challenge will testing (AI) enabled system is how (AI) performance can be measured. It would be NASA organization's difficulty to find (AI) technology to develop to carry on researching any space exploration missions in the future. However, (AI) technology will be a good economic benefit choice for space exploration mission in the future.

Why Developed And Developing Countries Need Artificial Intelligent Development To Assist Office Tasks

Must developed and developing countries need artificial intelligent development to assist office tasks ? Ought AI is needed to prefer to develop technique to assist office staffs to reduce workload to compare other kinds of occupation environment tasks aspects ? If one developed country, e.g. US, UK , Japan , Singapore it does not continue to develop artificial intelligence, robotic, then what disadvantges or weaknesses , it will encounter to compare when it chooses to continue to develop this artificial intelligent technology in society. If one developing country, e.g. China, Korea, Taiwan, it does not continue to develop artificial intelligence, robotic, then what disadantages or weaknesses, it will also encounter to compare when it chooses to continue to develop this artificial intelligent technology in in society. I shall explan the reasons why the results may cause to either the developed country, or the developing country as below:

● How AI help developing countries to communication and agriculture and learning and medical delivery development

Why can AI help developing countries ? Drones that pick inaccessible crops and mobile phones that give medical advice are two of the ways AI can transform life in the developing world. Artificial intelligence (AI) may improve the lives of the world's poor, the technology needed to revolutionise inefficient, ineffective food and healthcare systems in developing countries is well. For example, in low-income areas, agriculture and healthcare are two critical ecosystems that we can apply AI to immediately; this is not the far future, or even in five years.

Artificial intelligence (AI) has seeped into the daily lives of people in the developed world. From virtual assistants to recommendation engines, AI is in the news, our homes and offices. There is a lot of potential in terms of AI usage, especially in humanitarian areas. The impact could have a multiplier effect in developing countries, where resources are limited.

Emergency Response to developing countries' earthquake natural damage suddence occurrence predicting

AI and machine learning are still finding importance in emerging markets, but certain applications have emerged and are now widely used. For instance, predictive models for disaster relief enable first responders to automatically analyze large-scale behavior and movement through multiple sources of data including social media platforms, web forums, news sources, etc. Based on collected data, responders can scale reconstruction efforts and distribute supplies in a timely manner.

Why and how AI can assist farmers to predict when the earthquake occurs suddenly in order to avoid or reduce the natural damage to their agriculture productive number loss. For example, In 2015, when a major earthquake hit Nepal, more than 8 million people were affected. During the aftermath, drones were used to map and assess the destruction and speed up the rescue mission. The town of Sankhu, situated about 20 kilometers northeast of Kathmandu, was among the highly affected locations. In May 2018, my company Fusemachines and GeoSpatial Systems partnered with Sankhu's city officials to use drones and artificial intelligence in an effort to automatically estimate the reconstruction need. After processing data accumulated from a drone-powered aerial mapping of the region, the team fed this data to advanced machine learning algorithms. Combining drone imagery, digital mapping and machine learning, the team configured region modeling and infrastructure development with higher accuracy. Another organization known as One Concern, a California-based startup, has created a predictive AI program called Seismic Concern to accurately predict seism and is also working on solutions for wildfires, floods and hurricanes.

Smart AI Agriculture

Another application of AI in developing countries is smart agriculture. Farmers monitor crops more effectively and make better predictions on planting, weeding and harvesting using AI tools. It can also be used to analyze one plant at a time and add pesticides only to infected plants and trees instead of spraying pesticides across large swaths of crops. One California-based tech company is an example of this use of AI. So, the developing countries farmers in rural parts of India are also using AI to increase yields through better access to information about the farming season than they would normally have. Technology-enabled process automation offers the agribusiness industry the chance for remarkable growth -- not only in developed countries but around the world. There's a unique opportunity to increase yields, cut down labor costs and improve people's health.

Medicine Delivery to developing countries' patients urgent need

Companies are also leveraging AI to improve access to health care in some of the most remote areas of the world. In Rwanda, for example, Zipline is using drones to deliver medical supplies and blood to hospitals and clinics that are difficult to access by car. This has dramatically impacted people living in remote parts of the country because they are able to get medical help when needed. The drone system in Rwanda has also helped reduce waste of blood by 95%, as noted by Zipline. One Concern has created an AI program called Seismic Concern that accurately predicts seismic events and is also working on solutions for floods, wildfires and hurricanes. The medical field may actually benefit the most from emerging technologies in developing countries.

Assistance to reduce teaching work workload or psychological pressure to teachers in developing countries‘ schools

Another vital area benefiting from innovative technologies like AI is education. Advanced technologies can enhance how we learn, teach and perform tasks. In most developing countries, schools lack experienced teachers and resources to enhance students' knowledge. As a result, many students still have to walk long distances to get to the nearest school, which has created education gaps, especially in rural areas. AI tools such as personalized learning assistants can simplify learning by making tutoring services and learning materials accessible to all students, wherever they are. Machines can be automated to help students learn basic concepts without a tutor, which companies like Carnegie Learning are working on. This would allow students to learn at any time from anywhere. With AI, education is made easy and accessible to more people.

The initial usage of AI in developing countries has been at a micro level -- solving small, specific problems in a defined industry. As machine learning advances and there is a higher utilization of AI, we will see more complex issues being targeted and resolved. When duly adopted, AI can positively impact future developing countries people everyday lives not just in disaster intervention, education, health care and agriculture but can also help in mitigating poverty, malnutrition and pollution. Especially, in developing nations, to leverage AI's true potential and create a snowball effect. Startups are defining a holistic and humanitarian approach to building more sophisticated, AI-ready societies. Stakeholders in the AI landscape should understand the strengths and nuances of the developing world as well as the limitations of AI and create localized solutions and applications.

Why does smart phone help developing countries communication ?

Internet Seen as Positive Influence on Education but Negative on Morality in Emerging and Developing Nations. Internet access differs substantially across the 32 emerging and developing countries polled, with the lowest rates of internet use in South Asian and sub-Saharan African nations. Within countries, computer owners, young people, the well-educated, the wealthy and those with English language ability are much more likely to access the internet than their counterparts. To access the internet, people increasingly use smartphones rather than more cumbersome fixed landline connections and computers. Around the world, both smartphones and basic-feature phones alike are used for sending messages and taking pictures.

In fact, many developing countries young people, students are popular to use smart phones for internet usage aim, instead of communication. Moreover, many developing countries working people are also popular to use smart phones for any working usage in their working time , even non working time any time. So, smart phones (AI) phones will be important communication or leisure tools to developing countries people in the future. Unless, it is one day, scientists can develop another new communication tool to replace smart phones. So, artificial intelligence will be important to influence developing countries people , how to improve or bring positive learning attitudes to students in their daily learnnng lifes. as well as how to raise developing countries people, how to raise working people efficiency or improve performace in their daily working lifes. So, AI may bring positive learning or working attitudes to developing countries working people and students both.

The Positive Impact of Mass Media in Developing Countries

Radio, newspapers, television, Internet, social media, etc., all of these are forms of mass media. Each of these outlets has the capability of bringing information to thousands of people with one device. While in some communities it is easy to take advantage of these communication outlets such as television and Internet access, not everyone has access to such outlets. Radio is one of the most common forms of mass media in developing countries because it's affordable and uses less electricity than many other forms of mass media, but only approximately 75 percent of people in developing countries have access to a radio, and roughly 77 percent of people in rural areas have access to electricity.

For developing countries that have implemented forms of mass media in their communities, there have been numerous positive outcomes are influenced to impact developing countries mass media by artificial intelligence as below:

When AI is participated to developing countries mass media, it can influence any radio, television audiences raise more attention to each other through social media platforms such as Facebook and Twitter and create, organize and initiate street protests and campaigns. Furthermore, having access to social media in developing countries, people are able to connect to those that they usually wouldn't have the chance to talk to. Moreover, AI Provides educational opportunities- In many countries, the division between local and national languages as well as issues of literacy can make communication difficult. With the use of mass media, a bridge can be built between these two gaps. In India, there is a radio station that provides information in local languages and respects local culture and traditions. One of the main ways is to create public awareness of what is going on with businesses and government officials. The media plays an important role in giving people the opportunity to act against injustice, oppression and misdeeds that they otherwise wouldn't know about. Information on available healthcare, a mass radio broadcast was sent out encouraging parents to seek treatment at local healthcare facilities for their sick children. With this mass outreach on healthcare, the encouragement of people to take their children to healthcare facilities saved thousands of lives. This easy way of encouraging others and bringing awareness about certain diseases was made possible through a simple radio broadcast. Finally, when AI is particiapted to media, it may bring many social issues to life that otherwise would remain unknown to many people. In developing countries and communities like Burkina Faso, when the radio broadcast was released about malaria, diarrhea and pneumonia, people were educated and moved to action and knew to take their children to healthcare facilities for preventative care. As it is seen, having access to different media outlets is vital for those in developing countries. Here are three ways that those in developing countries can implement mass media to help their people and communities.

When AI is participated to any internet radio or internet newspaper mass online listening or reading channel. It can provide online radios or newspapers in public places- By providing online radios and newspapers in public areas it gives community members to access news, information and emergency warnings. Even though radios can be on the cheaper side, there are still many people that can't afford to have a radio in their home. By providing one in a local place, not only would it better educate the community members but also it will bring the community together. So, it can make media outlets a two-way platform- Creating a two-way platform between the community and those who are behind the radio stations, newspapers or broadcasts makes the community feel involved and that their voices are being heard. An organization called Soul City in sub-Saharan Africa is showing how well two-way platforms work by engaging their listeners and having them contribute thoughts and ideas about complex issues. Because developing countries radio listening audiences or newspaper readers are popular to accept computer online radio listening channel or online newspaper reading channel to replace traditional paper newspapers or radio machines. So, AI may raise their listening news or reading news leisure feeling from online mass media channel in the future.

● Why do developed countries need to develop AI
Artificial intelligence, or AI, is driving massive shifts across the globe, and every day more questions arise. What impact will AI have on the workforce and how can we prepare for it? How can we encourage economy-boosting and job-creating technologies? How can we ensure that AI will be implemented ethically and with minimal bias? How will society benefit? For developed country, such as US example. None of the US, Israel and Russia have a formal national AI policy yet. Private sector companies such as Google, Amazon and Apple and the US department of defence are driving the bulk of AI investment in the United States. Though Israel does not have a specific policy, it is keenly focused on AI and has seen the number of AI start-ups triple since 2014.
Developed country may learn whether what weakness it is lacking when it does not continue to develop AI from one another developed country. Which countries are approaching AI most effectively, and to what degree is there opportunity for greater international collaboration? It may be too early to tell; however, when analyzing the best practices of existing national AI policies, there is much that can be learned. These are the specific areas to consider. When one developed country continue to develop or research AI, it may bring these benefits as below:
On gathering Data aspect, from self-driving vehicles to smart cities, data is the driver behind AI. Innovation in the United States is limited without a national strategy that answers questions about protocol and ownership. France and Denmark, on the other hand, are opening government data. France is hosting troves of centrally collected public and private data that it plans to make available as part of its strategy. Conversely, by taking a restrictive position on issues of data collection (as indicated by the

implementation of General Data Protection Regulation), the EU is putting manufacturers and software designers at a disadvantage while balancing the demand for privacy. On raising technologica talent aspect, the demand for AI talent far outweighs the available supply. As a result, almost every nation's strategy addresses talent development. Canada's AI strategy is distinct in that it primarily focuses on research and talent strategy. The country boasts AI degree programmes and is building a $127 million research facility in Toronto. Companies like Facebook and my own company, Uptake, are investing in Canada to access this talent pool. On AI legal technological innovation aspect, a whole host of legal questions swirl around AI. The country is developing a bill for AI liability that will be ready in March 2019. The government hopes the legal framework will attract investors by providing a simple, comprehensive guideline to enable the broad use of AI systems. So, when the developed country applied AI technology to assist any lawyers to work, then AI can help them to reduce the workload to draft any legal documents more easier. So, any developed countries lawyers' draft legal documents time must reduce if the developed countries lawyers accept to apply AI to assist their legal works. One of the great promises of AI is its potential for improving quality of life. But without the right planning and oversight, we risk exacerbating problems of inequality or marginalizing groups of people. As an example, India's AI strategy is focused on leveraging the technology not only for economic growth, but also for social inclusion.

AI may bring what benefits to developed countries

From SIRI to self-driving cars, artificial intelligence (AI) is progressing rapidly. While science fiction often portrays AI as robots with human-like characteristics, AI can encompass anything from Google's search algorithms to IBM's Watson to autonomous weapons. Artificial intelligence today is properly known as narrow AI (or weak AI), in that it is designed to perform a narrow task (e.g. only facial recognition or only internet searches or only driving a car). However, the long-term goal of many researchers is to create general AI (AGI or strong AI). While narrow AI may outperform humans at whatever its specific task is, like playing chess or solving equations, AGI would outperform humans at nearly every cognitive task.

Why research AI safety? Would AI bring war when AI is continued to develop by developed countries? In the near term, the goal of keeping AI's impact on society beneficial motivates research in many areas, from economics and law to technical topics such as verification, validity, security and control. Whereas it may be little more than a minor nuisance if your laptop crashes or gets hacked, it becomes all the more important that an AI system does what you want it to do if it controls your car, your airplane, your pacemaker, your automated trading system or your power grid. Another short-term challenge is preventing a devastating arms race in lethal autonomous weapons.

In the long term, an important question is what will happen if the quest for strong AI succeeds and an AI system becomes better than humans at all cognitive tasks. As pointed out by I.J. Good in 1965, designing smarter AI systems is itself a cognitive task. Such a system could potentially undergo recursive self-improvement, triggering an intelligence explosion leaving human intellect far behind. By inventing revolutionary new technologies, such a superintelligence might help us eradicate war, disease, and poverty, and so the creation of strong AI might be the biggest event in human history. Some experts have expressed concern, though, that it might also be the last, unless we learn to align the goals of the AI with ours before it becomes superintelligent.

There are some who question whether strong AI will ever be achieved, and others who insist that the creation of superintelligent AI is guaranteed to be beneficial. At FLI we recognize both of these possibilities, but also recognize the potential for an artificial intelligence system to intentionally or unintentionally cause great harm. We believe research today will help us better prepare for and prevent such potentially negative consequences in the future, thus enjoying the benefits of AI while avoiding pitfalls.

How can AI be dangerous when developed countries continue to develop AI to become weapon to replace soldiers?

Most researchers agree that a superintelligent AI is unlikely to exhibit human emotions like love or hate, and that there is no reason to expect AI to become intentionally benevolent or malevolent. Instead, when considering how AI might become a risk, experts think two scenarios most likely:

The AI is programmed to do something devastating: Autonomous weapons are artificial intelligence systems that are programmed to kill. In the hands of the wrong person, these weapons could easily cause mass casualties. Moreover, an AI arms race could inadvertently lead to an AI war that also results in mass casualties. To avoid being thwarted by the enemy, these weapons would be designed to be extremely difficult to simply "turn off," so humans could plausibly lose control of such a situation. This risk is one that's present even with narrow AI, but grows as levels of AI intelligence and autonomy increase.

The AI is programmed to do something beneficial, but it develops a destructive method for achieving its goal: This can happen whenever we fail to fully align the AI's goals with ours, which is strikingly difficult. If you ask an obedient intelligent car to take you to the airport as fast as possible, it might get you there chased by helicopters and covered in vomit, doing not what you wanted but literally what you asked for. If a superintelligent system is tasked with a ambitious geoengineering project, it might wreak havoc with our ecosystem as a side effect, and view human attempts to stop it as a threat to be met. So, a super-intelligent AI will be extremely good at accomplishing its goals, and if those goals aren't aligned with ours, we have a problem. You're probably not an evil ant-hater who steps on ants out of malice, but if you're in charge of a hydroelectric green energy project and there's an anthill in the region to be flooded, too bad for the ants. A key goal of AI safety research is to never place humanity in the position of those ants.

Why the recent interest in AI safety ?

Stephen Hawking, Elon Musk, Steve Wozniak, Bill Gates, and many other big names in science and technology have recently expressed concern in the media and via open letters about the risks posed by AI, joined by many leading AI researchers. The idea that the quest for strong AI would ultimately succeed was long thought of as science fiction, centuries or more away. However, thanks to recent breakthroughs, many AI milestones, which experts viewed as decades away merely five years ago, have now been reached, making many experts take seriously the possibility of superintelligence in our lifetime. While some experts still guess that human-level AI is centuries away, most AI researches at the 2015 Puerto Rico Conference guessed that it would happen before 2060. Since it may take decades to complete the required safety research, it is prudent to start it now.

Because AI has the potential to become more intelligent than any human, we have no surprise way of predicting how it will behave. We can't use past technological developments as much of a basis because we've never created anything that has the ability to, wittingly or unwittingly, outsmart us. The best example of what we could face may be our own evolution. People now control the planet, not because we're the strongest, fastest or biggest, but because we're the smartest. If we're no longer the smartest, are we assured to remain in control?

A captivating conversation is taking place about the future of artificial intelligence and what it will/should mean for humanity. There are fascinating controversies where the world's leading experts disagree, such as: AI's future impact on the job market; if/when human-level AI will be developed; whether this will lead to an intelligence explosion; and whether this is something we should welcome or fear. But there are also many examples of of boring pseudo-controversies caused by people misunderstanding and talking past each other. When one developed country continue to develop AI, can itself country's all factories workers will lose jobs, due to AI can replace them to do simple works in factories, or any public transport drivers, e.g. bus drivers, ferry , tram, train drivers, they will lose jobs, when AI ( non manual driving drivers ) can replace all public transport drivers. So, some occupations will lose if developed countries continue to develop or research AI to replace human to do some simple jobs, such as some cooking jobs can be done by AI. So, it is possible that future cookers won't be needed, because AI cooking skills may be better than them to cook any good taste chinese or western food in restaurants. If you drive down the road, you have a subjective experience of colors, sounds, etc. But does a self-driving car have a subjective experience? Does it feel like anything at all to be a self-driving car? Although this mystery of consciousness is interesting in its own right, it's irrelevant to AI risk. If you get struck by a driverless car, it makes no difference to you whether it subjectively feels conscious. In the same way, what will affect us humans is what superintelligent AI does, not how it subjectively feels.

In fact, AI may be make any brokers jobs in financial market. the main concern of the beneficial-AI movement isn't with robots but with intelligence itself: specifically, intelligence whose goals are misaligned with ours. To cause us trouble, such misaligned superhuman intelligence needs no robotic body, merely an internet connection – this may enable outsmarting financial markets, out-inventing human researchers, out-manipulating human leaders, and developing weapons we cannot even understand. Even if building robots were physically impossible, a super-intelligent and super-wealthy AI could easily pay or manipulate many humans to unwittingly do its bidding. So, future brokers will be replaced by AI, when AI can be made to own financial brokers' analytical mind to make more accurate whether the share price will rise up or fall down to compare human financial brokers' analytical mind. The robot misconception is related to the myth that machines can't control humans. Intelligence enables control: humans control tigers not because we are stronger, but because we are smarter. This means that if we cede our position as smartest on our planet, it's possible that we might also cede control.

Not wasting time on the above-mentioned misconceptions lets us focus on true and interesting controversies where even the experts disagree. What sort of future do you want? Should we develop lethal autonomous weapons? What would you like to happen with job automation? What career advice would you give today's kids? Do you prefer new jobs replacing the old ones, or a jobless society where everyone enjoys a life of leisure and machine-produced wealth? Further down the road, would you like us to create superintelligent life and spread it through our cosmos? Will we control intelligent machines or will they control us? Will intelligent machines replace us, coexist with us, or merge with us? What will it mean to be human in the age of artificial intelligence?

Why do developed countries people need AI ?

Why do we assume that AI will require more and more physical space and more power when human intelligence continuously manages to miniaturize and reduce power consumption of its devices. How low the power needs and how small will the machines be by the time quantum computing becomes reality? Why do we assume that AI will exist as independent machines? If so, and the AI is able to improve its Intelligence by reprogramming itself, will machines driven by slower processors feel threatened, not by mere stupid humans, but by machines with faster processors? What would drive machines to reproduce themselves when there is no biological incentive, pressure or need to do so?

Who says superior AI will need or want to have a physical existence when an immaterial AI could evolve and preserve itself better from external dangers. What will happen if AI developed by competing ideologies, liberalism vs communism, reach maturity at the same time, will they fight for hegemony by trying to destroy each other physically and/or virtually. If AI is programmed to believe in God, and competing AI emerges programmed by muslims, christians or jews, how are the different AI's going to make sense of the different religious beliefs, are we going to have AI religious wars? What if the "powers that be" greatest fear is the emergence of a super AI that police's and rationalizes the distribution of wealth and food. A friendly super AI that is programmed to help humanity by, enforcing the declaration of Human Rights (the US is the only industrialized country that to this day has not signed this declaration) ending corruption and racism and protecting the environment.Most benefits of civilization stem from intelligence, so how can we enhance these benefits with artificial intelligence without being replaced on the job market and perhaps altogether?

Key to the process of machine learning are neural networks. These are brain-inspired networks of interconnected layers of algorithms, called neurons, that feed data into each other, and which can be trained to carry out specific tasks by modifying the importance attributed to input data as it passes between the layers. During training of these neural networks, the weights attached to different inputs will continue to be varied until the output from the neural network is very close to what is desired, at which point the network will have 'learned' how to carry out a particular task. A subset of machine learning is deep learning, where neural networks are expanded into sprawling networks with a huge number of layers that are trained using massive amounts of data. It is these deep neural networks that have fuelled the current leap forward in the ability of computers to carry out task like speech recognition and computer vision.

In conclusion, when developed countries continue to develop AI, it may bring positive advantages to bring raising productivies, or efficiencies, but it may also raise unemployment ratio to any low skill or low knowledge jobs in ther societies. However, human future society will need to change to be better to raise our living standard. But AI is one kind the best choice tool to achieve this aim in our future, so I agree developed countries continue to develop or research AI to be the super -human machine.

Artificial Intelligence Worker Brings

Working Environment Influences

Robots were once known only for the manufacturing business but today they are very much part of many workplaces. The future is even more promising for this wonder of artificial intelligence.Imagine a robot doing some of the major tasks of managers like using data to evaluate problems, making better decisions, monitoring team performance, and even setting goals.

Technology is playing a pivotal role in helping humans work more effectively. Since automation has become an integral part of business operations, we can predict that robots are soon going to replace many jobs that are today performed by humans. Now that the corporate world is also on the cusp of entering the robotic age, let's see what pros and cons this technology offers business world. If one day, our global working environments have any kinds of robotic participates to our service and warehouse and office etc. different working environment in order to assist office workers, service workers, warehouse workers, professional lawyers, doctors accountants job duties, what positive or negative influences, it will bring to what negative or positive effects to any office , warehouse, shopping centre, hospital, transport , restaurant etc, different working environments. Can robotic help office , warehouse to raise efficency ? Can robotic help hospital, restaurant, cinema, shopping center to improve service performance? Can robotic influence working environment to be worse? Can robotic help office or any working places to reduce expenditure or reduce long time machine and salary cost when they do not need more employees or machines , due to robotic workers assistance.

I shall attempt to explain whether robotic workers will bring what positive or negative influence to our future working environment as below:

Advantages to robotic bring to working environment

What advantages thar robotic will bring to working environment? They may include: Many people fear that robots or full automation may someday take their jobs, but this is simply not the case. Robots bring more advantages than disadvantages to the workplace. They enrich a company's ability to succeed while improving the lives of real, human employees who are still needed to keep operations running smoothly. If you're thinking about investing in some robots, share the advantages with your employees. You might be surprised at how many of them are quick to support the idea.

1. Safety

Safety is the most obvious advantage of utilizing robotics. Heavy machinery, machinery that runs at hot temperature, and sharp objects can easily injure a human being. By delegating dangerous tasks to a robot, you're more likely to look at a repair bill than a serious medical bill or a lawsuit. Employees who work dangerous jobs will be thankful that robots can remove some of the risks.

2. Speed

Robots don't get distracted or need to take breaks. They don't request vacation time or ask to leave an hour early. A robot will never feel stressed out and start running slower. They also don't need to be invited to employee meetings or training session. Robots can work all the time, and this speeds up production. They keep your employees from having to overwork themselves to meet high pressure deadlines or seemingly impossible standards.

3. Consistency

Robots never need to divide their attention between a multitude of things. Their work is never contingent on the work of other people. They won't have unexpected emergencies, and they won't need to be relocated to complete a different time sensitive task. They're always there, and they're doing what they're supposed to do. Automation is typically far more reliable than human labor.

4. Perfection

Robots will always deliver quality. Since they're programmed for precise, repetitive motion, they're less likely to make mistakes. In some ways, robots are simultaneously an employee and a quality control system. A lack of quirks and preferences, combined with the eliminated possibility of human error, will create a predictably perfect product every time.

5. Happier Employees

Since robots are often assigned to perform tasks that people don't particularly enjoy, like menial work, repetitive motion, or dangerous jobs, your employees are more likely to be happy. They'll be focusing on more engaging work that's less likely to grind down their nerves. They might want to take advantage of additional educational opportunities, utilize your employee wellness program, or participate in an innovative workplace project. They'll be happy to let the robots do the work that leaves them feeling burned out.

6. Job Creation

Robots don't take jobs away. They merely change the jobs that exist. Robots need people for monitoring and supervision. The more robots we need, the more people we'll need to build those robots. By training your employees to work with robots, you're giving them a reason to stay motivated in their position with your company. They'll be there for the advancements and they'll have the unique opportunity to develop a new set of tech or engineering related skills.

7. Productivity

Robots can't do everything. Some jobs absolutely need to be completed by a human. If your human employees aren't caught up doing the things that could have easily be left for robots, they'll be available and productive. They can talk to customers, answer emails and social media comments, help with branding and marketing, and sell products. You'll be amazed at how much they can accomplish when the grunt work isn't weighing them down.

8. Cost reduce

The first and the foremost advantage of having robots in workplaces is their cost. Robots are much cheaper than humans and their cost is now decreasing. It's a fact that we cannot compare human abilities with robots but robotic capabilities are now growing quickly. For example, if you run an essay writing service, you can use robots to perform every kind of research related to any subject. Because robots are more active and don't get tired like humans, the collaboration between humans and robots is reducing absenteeism. The pace of human cannot increase hence robots are helping humans.

However,robots are more precise than humans; they don't tremble or shake as human hands. Robots have smaller and versatile moving parts which help them in performing tasks with more accuracy than humans. There is no doubt that robots are significantly stronger and faster than humans. Robots come in any shape and size, depending upon the need of the task. Robots can work anywhere in any environmental condition whether it is space, underwater, in extreme heat or wind etc. Robots can be used everywhere where human safety is a huge concern. Robots are programmed by a human; they cannot say no to anything and can be used for any dangerous and unwanted work where humans may deny to offer their services. For example, many robotic probes have been sent into space but have never returned. Robots in warfare are saving more lives and have now proven to be very successful. For example, in chemical factory environment, robots are now being used in the chemical industry and can, for example deal with chemical spills in a nuclear plant, which would otherwise pose a major health concern. Cost-effectiveness is one of the most sound arguments to be made for the case of industrial robots. Robots will reduce production costs by eliminating internal costs to compensate human salaries. Businesses are forecasting that their profitability will increase once they implement robots into production, or that they will have more financial mobility to invest in new products or technologies.

9. productive efficiency

Quality assurance is expected with the use of machinery in production. Industrial robots will be able to ensure consistency with mass production of manufactured products. The possible human error that assembly line workers pose the threat of will be removed. Optimized production efficiency means that a general manager will be able to have set quantity and quality standards that will be met by robots. Production quotas will not be jeopardized by low concentration, break time and employee injuries, among other things. The efficiency of production forecasts and supply levels will be increased with robots, able to be programmed to work at the optimal speed for a given plant. Limiting human work in hazardous environments, because manufacturing jobs often place workers at more physical risk compared to a lot of other industries. Lowering the level of a hazard presented to employees on the job is attractive to executives to preserve company reputation and minimize potential legal liabilities.

10. Reducing longer working hours

Typically people have to have breaks, get distracted and after time attention drops and pace slows. With a robot it can work 24/7, and keeps running at 100%. Typically if you replace one person on a key process in a production line with a robot the output increases by 40% in the same working hours just because a robot has more stamina and never stops. Robots also don't take holidays or have unexpected days off sick.

11. Increased profitability

By increasing the efficiency of your production process, reducing the resource and time needed to complete it, and also achieving higher quality products, industrial robots can thus be used to achieve higher profitability levels overall, with lower cost per product.

12. Improved working environment

Industrial robots are often used for performing tasks which are deemed as dangerous for humans, as well as being able to perform highly laborious and repetitive tasks. Overall, by using industrial robots you can improve the working conditions and safety in your factory or production process. Robots don't get tired and make dangerous mistakes, neither do they suffer from repetitive strain injury.Due to their high accuracy levels, robots can also be used to produce higher quality products which adhere to certain standards of quality, whilst also reducing the time needed for quality control.Industrial robots are able to complete certain tasks faster and better than people, as they are designed to perform these tasks with a higher accuracy level. This and the fact that they are used to automate processes which previously might have taken significantly more time and resources, means that you can often use industrial robots to increase the efficiency of your production line.

13. Improved Quality Assurance

Few workers enjoy doing repetitive tasks and after a certain period of time concentration levels will naturally decline. This lapse in concentration is known as vigilance decrement and can often lead to costly errors for the business and sometimes serious injury to the member of staff.Robotic automation eliminates these risks by accurately producing and checking items meet the required standard without fail. With more product going out the door manufactured to a higher standard, this creates a number of new business possibilities for companies to expand upon.

14. Increased Productivity

Using robotic automation to tackle repetitive tasks makes complete sense. Robots are designed to make repetitive movements. Humans, also by design, are not. The introduction of automation into your manufacturing process has many different productivity benefits, some of which are shown here.Giving staff members the opportunity to expand on their skills and work in other areas will create a better environment which the business as a whole will benefit from. With higher energy levels and more focus put into their work, the product can only improve, which will also lead to extremely satisfied clients.

15. Avoiding workers need to work In Hazardous Environments

Aside from potential injuries in the workplace, staff members in particular industries can be asked to work in unstable or dangerous environments. For example, if a high level of chemicals are present, robotic automation offers the ideal solution, as it will continue to work without harm. Production areas that require extremely high or low temperatures typically have a high turnover of staff due to the nature of the work. Automated robots can minimise material waste and remove the need for humans to put themselves at unnecessary risk.

Disadvantages to robotic bring to working environment

1. Increase unemployment rate and job loss

On working environment cost increasing aspect, where robots are increasing the efficiency in many businesses, they are also increasing the unemployment rate. Because of robots, human labour is no longer required in many factories and manufacturing plants. They can certainly handle their prescribed tasks, but they typically cannot handle unexpected situations.The ROI of your business may suffer if your operation relies on too many robots. They have higher expenses than humans, so at the end of the day you may not always achieve the desired ROI.

However, robots may have AI but they are certainly not as intelligent as humans. They can never improve their jobs outside the pre-defined programming because they simply cannot think for themselves. Robots installed in workplaces still require manual labour attached to them. Training those employees on how to work with the robots definitely has a cost attached to it.Moreover, robots have no sense of emotions or conscience. They lack empathy and this is one major disadvantage of having an emotionless workplace.Also, robots operate on the basis of information fed to them through a chip. If one thing goes wrong the entire company bears the loss. Where a robot saves times, on the other hand it can also result in a lag. It is, after all, a machine so you cannot expect too much from them. If a robot malfunctions, you need extra time to fix it, which would require reprogramming.If ultimately robots would do all the work, and the humans will just sit and monitor them, health hazards will increase rapidly. Obesity will be on top of the list. So there are advantages, but there are disadvantages as well. It is the twenty first century and we cannot work without machines.Humans are still considered far more efficient than robots when it comes to decision making powers, handling difficult situations, brainstorming, and generally bringing a sense of emotion and empathy into a workplace. Besides, you cannot rule out the significant role of humans in a business. After all, no machine can replace the human factor 'real employees' bring into a workplace. So, AI can raise unemployment and increase factory or shopping center or office working environment cost when their working environment are applied robotic to replace many workers, then machine electricity expense will also increase. Otherwise, human workers can not spend too much electricity expense in cost aspect.

Whilst industrial robots can prove highly effective and bring you a positive ROI, implementing them might require a fairly high capital cost. That's why, before making a decision we recommend considering both the investment needed and also the ROI you expect to achieve. Often the easiest way to get round this issue is to take out asset finance and the ROI of the robot more than pays for the interest on the asset finance.

This is typically the biggest obstacle that will decide whether or not a company will invest in robotic automation, or wait until a later stage. A comprehensive business case must be built when considering the implementation of this technology. The returns can be substantial and quite often occur within a short space of time. However, the cash flow must be sustainable in the meantime and the stability of the company is by no means worth the risk if the returns are only marginal. Yet, in most instances there will be a repayment schedule available, which makes it a lot easier to afford and control finances. Our downloadable automation payback calculator also has a finance scheme option so you can see how this would work for you.

On job loss increasing aspect, Job loss is by far the most significant opposition frequently brought against the use of robots in the manufacturing industry. Industry workers of all levels, from entry-level to veterans, worry about the security of their employment status, and the ability of their job to be replaced by a robot. This panic is more widespread in this industry compared to others because of the closer immanence of a robot takeover in manufacturing.

Macro effects are another topic that usually comes up with job loss. More "big picture" thinkers wonder how the national, and eventually global economy will be affected when manufacturing workers' jobs are displaced. How can this mass unemployment possibly be compensated for, and how can the robots' presumed success be limited from seeping into other industries. However, increased investment costs are a financial counterpoint to industrial robots, with the idea that manufacturing companies will rack up their debt investing in robotic technology. Firms that do not have the funding might even go bankrupt in an effort to keep up with industry trends rather than continue on with normalized operations.Hence, elimination of a whole labor class would presumably occur a bit of a ways down the road, but the implications of this point are too large not to consider. Bringing in robots to take unskilled labor jobs will place more pressure on the economy, education system, and financial market, just to name a few. The United States has always been associated with the grit and work ethic of its blue-collar workers, and robots are threatening to eliminate this aspect of the human population, with a take over of production jobs.

One of the biggest concerns surrounding the introduction of robotic automation is the impact of jobs for workers. If a robot can perform at a faster, more consistent rate, then the fear is that humans may not be needed at all. While these worries are understandable, they are not really accurate.The same was said during the early years of the industrial revolution, and as history has showed us, humans continued to play an essential role. Amazon are a great example of this. The employment rate has grown rapidly during a period where they have gone from using around 1,000 robots to over 45,000

2. Robotic can not perform better to compare human workers, when they need to work long time in any working environment
Robots need a supply of power, The people can lose jobs in factories, They need maintenance to keep them running, It costs a lot of money to make or buy robots, The software and the equipment that you need to use with the robot cost much money. Robots cost much money in maintenance & repair, The programs need to be updated to suit the changing requirements, the machines need to be made smarter, In case of breakdown, the cost of repair may be very high, The procedures to restore lost code or data may be time-consuming & costly.
Robots can store large amounts of data but the storage, access, retrieval is not as effective as the human brain, They can perform repetitive tasks for a long time but they do not get better with experience such as the humans do. Robots are not able to act any different from what they are programmed to do, With the heavy application of robots, the humans may become overly dependent on the machines, losing their mental capacities, If the control of robots goes in the wrong hands, Robots may cause the destruction. Robots are not intelligent or sentient, They can never improve the results of their jobs outside of their predefined programming, They do not think, They do not have emotions or conscience, This limits how the robots can help & interact with people. Robots can take the place of many humans in factories, So, the people have to find new jobs or be retrained, They can take the place of the humans in several situations, If the robots begin to replace the humans in every field, They will lead to unemployment.
Humans fear robots, Robots inspire two types of fear: firstly, that they might take over our jobs, and secondly, that they could take over the world, Robots will steal our jobs, Robots have the effect of increasing productivity rather than eliminating jobs.Robotics become increasingly present in our everyday life, with household robots, medical, industrial, on production lines, not to mention airports, banks, and hotels, So, Robots may dominate the human species. Robots can operate on the basis of information fed to them through a chip, when one thing goes wrong the entire company bears a loss.The robot can save times, but it can also result in a lag, It is a machine so you can't expect too much from them, If the robot has malfunctioned, you need extra time to fix it, which would require reprogramming, If robots would do all the work, and the humans will just sit and monitor them, health hazards will increase rapidly, Obesity will be on top of the list and less labour at workplaces.

3. Increasing training expense

Whilst industrial robots are excellent for performing many tasks, as with any other type of technology, they require more training and expertise to initially set up. The expertise of a good automation company with a support package will be very important. To minimise your reliance on automation companies you can train some of your engineers on how to program robots, but you will still need the assistance of experienced automation companies for the original integration of the robot.

In recent years the number of industrial robots and the applications they can be used for has increased significantly. However, there still are some limitations in terms of the type of tasks they can perform, which is why we suggest that an automation company looks at your requirement to assess the options first. Sometimes a bespoke automated system may give a better or faster result than a robot. Also, a robot does not have everything built into it, often the success or failure of an industrial robotic system depends on how well the surrounding systems are integrated e.g. grippers, vision systems, conveyor systems etc. Only use good trusted robot integrators to be sure of the optimum results if you do choose to use industrial robots.

ON conclusion, future robotic development life mature stage ought concentrate on researching how it can be applied to serve office workers to raise efficiency in order to help organizations reduce productivities cost expense in long time.

www.ingramcontent.com/pod-product-compliance
Ingram Content Group UK Ltd.
Pitfield, Milton Keynes, MK11 3LW, UK
UKHW041633190726
13854UKWH00006B/2469